The Fundamental Rules
of Risk Management

T0320281

CHAPMAN & HALL/CRC FINANCE SERIES

Series Editor

Michael K. Ong

Stuart School of Business
Illinois Institute of Technology
Chicago, Illinois, U. S. A.

Aims and Scopes

As the vast field of finance continues to rapidly expand, it becomes increasingly important to present the latest research and applications to academics, practitioners, and students in the field.

An active and timely forum for both traditional and modern developments in the financial sector, this finance series aims to promote the whole spectrum of traditional and classic disciplines in banking and money, general finance and investments (economics, econometrics, corporate finance and valuation, treasury management, and asset and liability management), mergers and acquisitions, insurance, tax and accounting, and compliance and regulatory issues. The series also captures new and modern developments in risk management (market risk, credit risk, operational risk, capital attribution, and liquidity risk), behavioral finance, trading and financial markets innovations, financial engineering, alternative investments and the hedge funds industry, and financial crisis management.

The series will consider a broad range of textbooks, reference works, and handbooks that appeal to academics, practitioners, and students. The inclusion of numerical code and concrete real-world case studies is highly encouraged.

Published Titles

Decision Options®: The Art and Science of Making Decisions, **Gill Eapen**

Emerging Markets: Performance, Analysis, and Innovation, **Greg N. Gregoriou**

Introduction to Financial Models for Management and Planning, **James R. Morris and John P. Daley**

Pension Fund Risk Management: Financial and Actuarial Modeling, **Marco Micocci, Greg N. Gregoriou, and Giovanni Batista Masala**

Stock Market Volatility, **Greg N. Gregoriou**

Portfolio Optimization, **Michael J. Best**

Operational Risk Modelling and Management, **Claudio Franzetti**

Handbook of Solvency for Actuaries and Risk Managers: Theory and Practice, **Arne Sandström**

Quantitative Operational Risk Models, **Catalina Bolancé, Montserrat Guillén, Jim Gustafsson, Jens Perch Nielsen**

The Fundamental Rules of Risk Management, **Nigel Da Costa Lewis**

Proposals for the series should be submitted to the series editor above or directly to:
CRC Press, Taylor & Francis Group
4th, Floor, Albert House
1-4 Singer Street
London EC2A 4BQ
UK

CHAPMAN & HALL/CRC FINANCE SERIES

The Fundamental Rules of Risk Management

Nigel Da Costa Lewis

Managing Director, AusCov

CRC Press
Taylor & Francis Group
Boca Raton London New York

CRC Press is an imprint of the
Taylor & Francis Group, an **informa** business
A CHAPMAN & HALL BOOK

CRC Press
Taylor & Francis Group
6000 Broken Sound Parkway NW, Suite 300
Boca Raton, FL 33487-2742

First issued in paperback 2019

ISBN-13: 978-1-4398-1618-9 (hbk)
ISBN-13: 978-0-367-38131-8 (pbk)

This book contains information obtained from authentic and highly regarded sources. Reasonable efforts have been made to publish reliable data and information, but the author and publisher cannot assume responsibility for the validity of all materials or the consequences of their use. The authors and publishers have attempted to trace the copyright holders of all material reproduced in this publication and apologize to copyright holders if permission to publish in this form has not been obtained. If any copyright material has not been acknowledged please write and let us know so we may rectify in any future reprint.

Library of Congress Cataloging-in-Publication Data

Lewis, Nigel Da Costa.
  The fundamental rules of risk management / Nigel Da Costa Lewis.
    p. cm. -- (Chapman & Hall/CRC finance series)
  Includes bibliographical references and index.
  ISBN 978-1-4398-1618-9 (hardback)
    1. Risk management. 2. Risk management--Case studies. 3. Financial risk management. 4. Financial risk management--Case studies. I. Lewis, Nigel Da Costa. II. Title.

HD61.L486 2012
658.15′5--dc23   2012013466

Visit the Taylor & Francis Web site at
http://www.taylorandfrancis.com

and the CRC Press Web site at
http://www.crcpress.com

To Denise, my sister, your smile lifts the hearts of those who know you.

Your voice spurs our gaggle of youngsters onto higher ground.

Contents

Preface...xi

Section I   The Behavioral Foundations of Risk Management

1. Unreason Is the Even Eviler Twin Brother of Greed..............................3
   A Word to the Wise—You Cannot Rely on the Flynn Effect...................7
   The Unintended Consequences of the Glad Game...........................12
   But You Have to Remember Ivar Kreuger of Kalmar!............................17
   Endnotes...26

2. The Maleficent Hand of the Men in Gray Suits.....................................29
   Unreason Abounds in Places Where It Must Not......................................29
   The Conspiratorial Regulator..31
      The Apathetic Regulator 43
   Endnotes...50

3. The Unpalatable Truth about Risk Management 53
   A Rather Vulgar, But Common, Perception of Risk Management 54
   The Emperor of Risk, His Lyre and the Palatine................................57
   The Utter and Total Redundancy of Financial Risk Management.........58
   The Risk Manager as a "Quivering Dastard" 59
   Perception and Reality about Risk Management..............................62
   For Further Thought...65
   Additional Resources 66
   Appendix 67
   Endnotes...68

Section II   What You Need to Know, But Nobody Wants to Tell You

4. What the Textbooks Will Not Tell You about Corporate
   Governance 73
   The Essence of the Governance Issue 76
   The Superficiality of Compliance 77
   Why "Gentleman's" Agreements Do Not Work................................80
   The Role of Criminal Penalty...83

The Benefit of Wolf Pack Capitalism 86
The Inherent Ethos of Risk Management 88
The Cost of Corporate Governance 93
Why Governance Failures Are Inevitable 95
For Further Thought 99
Additional Resources 102
Endnotes 105

5. The Most Important Lesson a Risk Manager Must Know 111
Odysseus and the Sirens' Song 114
The Consequence of Ignoring the Golden Rule 115
An Immutable Condition for Success in Risk Management 117
For Further Thought 119
Additional Resources 120
Endnotes 121

6. A Powerful Secret from Henry Fayol 123
The Great Work: General and Industrial Management 126
The Rise of Fayol's "Strategic Security Director" 127
The Warren Buffet Principle of Risk Management 129
Can Chief Risk Officers Add Value? 131
For Further Thought 133
Additional Resources 134
Endnotes 136

7. The Incredible Advantage of a Monocle on Risk 139
What Is a Monocle on Risk? 140
The Hidden Dangers of Risk Management Silos 141
The Need for Better Risk Management 143
The Challenge 144
The Three Essential Elements of Successful Risk Integration ... 146
For Further Thought 148
Additional Resources 152
Endnotes 153

8. Benefit from the Fable of Spreadsheet City 155
Don't Be a Victim of Spreadsheet Hell 157
Why Spreadsheet Failure Costs Big Time! 158
How to Bring Spreadsheet Risk under Control 162
Understanding the Nature of Spreadsheet Error 163
The Principles of Spreadsheet Engineering 164
The Potential of Compilable Spreadsheets 166
Seven Rules for Superior Spreadsheet Design 167
How to Minimize Risk through Formal Testing 169
For Further Thought 171

Additional Resources 173
Endnotes.. 174

**9. How to Guarantee Success by Understanding the Nature
of Failure**... 177
The Value Added of Vendor Risk Information Systems 178
How to Guarantee Success by Understanding the Nature
of Failure 179
Developing a Winning Game Plan 181
Creating a High Performance Team.. 182
The Important Lesson of $\frac{1}{2} \times n \times (n-1)$ 183
The Critical Role of Executive Buy-In 184
Clarifying Your Requirements.. 185
The Truth about Project Managers.. 189
For Further Thought.. 191
Additional Resources 194
Endnotes.. 196

10. Snake Oil Salesmen, Goat Gonads, and Value at Risk...................... 201
VaR Explained 202
The Joyous Exclamation of Simons 204
The Tipping Point 205
What the *Rocket Scientists* May Not Tell You, But You Need
to Know 208
The Curse of the Bell-Shaped Curve.. 209
Exact Imprecision—On the Accuracy of VaR 210
For Further Thought.. 212
Additional Resources 213
Endnotes.. 215

Index.. 219

Preface

Author's Warning

Knowledge of the fundamental rules of risk management is powerful. Accomplished business leaders and prosperous entrepreneurs the world over understand these rules intuitively. They exploit them continually as they rise to the very top of their industries. The material you are about to read is based on private knowledge gathered during the years since the publication of my three earlier texts. My first book, *Market Risk Modelling* (Risk Books, 2003), was released in 2003, and it became an immediate best seller. Since then, I have presented at numerous conferences where attendees paid many hundreds of dollars to be present. My ideas have been published in investment journals, and I have spoken at private seminars where select groups of successful professionals have shared their views. All the while, I have been working, writing, thinking, and observing risk. This book in your hands is an enlarged, revised, and updated edition of all my previous works on the subject.

Taking on risk is like building a bomb; when you ignore the fundamental rules, you bury it alive. It will explode. Maybe not today or tomorrow, but one day it will explode. The consequences, you will discover as you read the book, to personal finances, professional careers, corporate survivability, and even nation states, can be ruinous. The very strange thing about the topics discussed in this book is that they are not well known nor are they discussed in risk management circles. They do not form part of the curriculum by which risk managers are certified. Yet, comprehension of them is essential for success. Crystal clarity on risk separates successful professionals, companies, and economies from history's forgotten failures. Remember the investment house of Hornblower and Weeks? What about Lee, Higginson & Co.? I thought not!

Unfortunately, for nonquantitative types the discipline is dominated by mathematicians, econometricians, and statisticians. The old adage of *measure what you want and reward what you measure* provides a naturally appealing environment for the quantitatively inclined. Indeed, pick up almost any text on risk management and you will be confronted by an array of equations and probability distributions. However, I cannot stress enough that without clear knowledge and continuous application of the fundamental rules of risk management, quantification offers little more than a dangerous facade of precision and accuracy. Fortunately, the fundamental rules of risk management are easily explained. You do not require a Ph.D. in statistics from

Cambridge University to understand the pages you are about to read. You need nothing more than an eager interest.

There is only a limited amount of information that can be included in a book of this size. Additional material, publications, and resources can be found by visiting my Web site: www.NigelDLewis.com.

Nigel Da Costa Lewis

Section I

The Behavioral Foundations of Risk Management

We make no bones about what risk is in this text. Investors fear most an irreversible loss of wealth. *Risk is a permanent loss of capital.* We offer no other definition. In the first section of this book, we delve deep into its behavioral roots. Through careful illustration, we detail in historical and contemporary contexts the indubitable truths surrounding many of the behavioral biases, which induce risk. We expose the fallacy of the wisdom of experts, explain with crystal clarity why you cannot rely on regulators, outline the characteristics of the *glad game*, and demonstrate how high intelligence or lack thereof can ultimately act as a fetter over which hard-earned wealth can be spilt and ultimately evaporate into the pockets of others. We end this section of the book with a candid discussion of the weaknesses and failures of modern risk management. We face up to many of the unspoken *eight hundred pound gorillas* in the room. These include a rather vulgar but common perception of risk managers, the utter and total redundancy of financial risk management, and the lore surrounding the quivering dastards of risk mismanagement. Throughout the text, we hope the reader will keep in the front of their mind the incontrovertible truth that *risk realized is always personal.* It can deal a devastating life-altering blow and *risk management cannot therefore be taken lightly.*

1

Unreason Is the Even Eviler
Twin Brother of Greed

And together they have been an essential feature of every financial crisis, large and small, since history began. It was greed and unreason which drove forward the spectacular Wall Street Crash of 1929; greed and unreason which lay at the heart of the Japanese property bubble of the 1980s, and greed and unreason which foreshadowed the global financial crisis of 2008. Even today, many of our well-known corporations and financial institutions will have abbreviated lives principally due to greed and unreason. Many of our industrial leaders, even those who are titans of our time, will falter and fail principally due to greed and unreason. Very many of today's hard working, hard saving citizens will fail to garner enough wealth to enjoy a leisurely retirement, in large part because of greed and unreason. Corporations and financial institutions, industrial leaders, and hard working individuals who stumble at the hands of greed and unreason do so precisely because they are unaware when greed and unreason have come to dominance; they lack a clear understanding of the fundamental rules of risk management—and that is their main problem.

The point is made piercingly clear by the tulip madness, which gripped the Netherlands during its golden age. The tulip originated in the mountains of central Asia. As early as the year 1050 they were cultivated in Isfahan and Baghdad. The flower gradually made its way west via the Ottoman Empire as a prized object of the sultans. The first tulip recorded in Europe was seen in 1559 in the town of Augsburg, Germany and was imported directly from Constantinople.[1] The blossom rapidly caught on as an exotic flower being proudly displayed as the centerpiece of elaborate gardens and collections by wealthy individuals across the entire continent. However, it was in Holland in the early 17th century, which had entered a period of economic prosperity coinciding with the Eighty Years' War against Spain, where the Middle Eastern blossom took its firmest and most bizarre root.

To appreciate the significance, it is necessary to have a little historical background. In 1576, the great medieval town of Antwerp, then part of the Duchy of Brabant, was captured and sacked by the Spanish. Nine years later, the Spanish, under Alessandro Farnese, Duke of Parma and Piacenza, sacked the city again and all Protestant citizens were expelled. Thus, for many years, the people of the lowlands of Europe, with their many beautiful cities, majestic landscapes, and natural rural splendor were under the persistent scourge of Spain.

The fate of the lowlanders, modern day Netherlanders and Belgians, turned dramatically when Frederick Henry, the son of assassinated Willem de Zwijger, became Stadtholder and Prince of Orange. He was able to decisively repulse the Spanish and for more than a quarter century a great golden age of prosperity descended on Holland. Commerce expanded, the artisan class flourished, and some of the greatest painters and printmakers in European art history such as Rembrandt and Vermeer emerged. It was thus among the economic prosperity that had arisen in the flatlands of Europe where, according to the Scottish writer and historian Charles Mackay, the encomia lavished on the tulip root reached the highest peak.

In the wild, tulips are generally solid, bold colors; pink, purple, red, and the like. However, they are on occasion subject to the tulip breaking virus. The virus, first recorded by physician and botanist Charles de L'Écluse in 1576, causes colorful variegations in the petal of pink, purple, and red flowered tulips in addition to mottling of the leaves. So beautifully variegated are the petals that they are often known as *Rembrandt tulips* because they were favorite subjects in many paintings by the Dutch Masters.

The source of the variegation remained a complete mystery to Charles de L'Écluse and subsequent botanists well into the 20th century. It was initially thought to be caused by environmental factors. The most delicate variegation, it was suggested, could be induced by a combination of frequently changing the soil, allowing the bulb to seed, and storage of resting bulbs in an exposed position so that they could be "acted" upon by the natural elements—wind, rain, frost, and sun.

It was not until 1927, when the Englishwoman Dorothy Cayley working at the John Innes Horticultural Institution on the outskirts of London discovered the true cause. Her discovery took the botanic world by storm and propelled the term *mycological* into the lexicon of both amateur and professional tulip fanciers worldwide.

By the middle of the 19th century it was becoming increasingly clear to botanists and scientists that:

> Just below the earth's surface, hidden in the soil from which plants, trees, and grasses spring, lies a kingdom unlike any other—the kingdom of fungi. It's a family of life unto its own, one of the least explored and understood by modern science, and yet these creatures who number in the tens of thousands have as great a part to play in every ecosystem as any vegetable, vermin, or viper struggling in the web of life.[2]

The British Mycological Society, founded in 1896, drew together an eclectic mix of scientists interested in the developing discipline of fungal science. Dorothy Cayley became, for a time, the British Mycological Society's most well-known member; and in doing so, propelled mycology to the forefront of the physical sciences. You see, Dorothy had a most unusual fascination with plant diseases, soils, and in particular the slime mold Mycetozoa. It is reported that she *would sit up all night watching their growth and development under the microscope.*[3]

In 1910, Dorothy joined the John Innes Horticultural Institution as a volunteer worker. Within a few short years, she was offered a studentship and eventually promotion to the title Mycologist. It was her careful study of the diseases of peas and fruit, and the life history of "die-back" fungus, *Diaporthe pernicisiosa*, which laid the groundwork for her 1927 unearthing of the source of the variegated tulip.

In a series of quite remarkable experiments involving the transfer of infected tissue from "broken" bulbs to healthy bulbs, Dorothy discovered the infective agent that caused the variegation would also be transferred. It turned out that in the wild, the nonfatal infection was spread from tulip to tulip by several aphid species; *Myzus persica*, *Macrosiphum euphorbiae*, *Aphis fabae*, *Aphis gossypii*, *Dysaphis tulipae*, and *Aulocorthum circumflexum*. As the aphid bites into an infected plant, small amounts of the virus are transferred into the plant's vascular system. The result of this natural process is beautifully variegated petals.[4]

It was the delicate bars, stripes, streaks, featherings, and flames of different colors, which were highly prized by connoisseurs of the wealthy merchant class of 17th century Holland. With little knowledge of the cause of the variegation, demand for the rare blossoms was high. The Dutch tulip trade was thus begun.

During 1634, demand for variegated blossoms became so great in Holland that Mackay noted *the ordinary industry of the country was neglected, and the population, even to its lowest dregs, embarked in the tulip trade.* So widespread was the demand for tulips during the 1630s that a code of laws was drawn up to regulate the trade with the tulip notary replacing the public notary as the profession of choice in many towns. Regular trades took place on the Stock Exchange of Amsterdam, Rotterdam, Harlaem, Leyden, Alkmaar, and Hoorn. Market makers in traditional stocks turned their attention to making markets in tulips, and at great profit.

As word spread of the riches to be made, tulip speculation abounded and the price continued to rise and rise. So rapid was the price increase, so profitable the opportunity that tulip trading spread across the entire length and breadth of the country. As Mackay describes:

> A golden bait hung temptingly out before the people, and one after the other, they rushed to the tulip-marts, like flies around a honey-pot. (p. 69)

As with all great moments in human history, the events surrounding Dutch tulip madness are steeped in a combination of folklore and myth. One such frequently recounted tale involves an English gentleman, who, in the modern retelling and echoing the eventual cause of variegation, is described as more insect-like than human. The English gentleman had been traveling throughout the lowlands of Europe and happened upon a wealthy Dutchman. The Dutch to this day are an especially hospitable people, particularly to foreigners. It should come as no surprise therefore to learn the Englishman, himself being rather a rarity in Holland at that time, was invited to repose for a few days at the Dutchman's wealthy estate. Many hours of jocular conversation was had by all.

At some point the Englishman was left alone, perhaps during the time when men of that day retired to their rooms prior to dinner. Whatever the case, the gentleman found himself in the conservatory of his wealthy host. It was a huge brick and stone structure with tall glass windows, a solid beamed roof, and a cast iron stove heater. There on a large wooden table he spied a disheveled "onion-like" root. Being somewhat of an amateur botanist, he was drawn by its unusual ugliness. He picked it up in his right hand and examined it closely. Now, one must remember that English gentlemen of that day always had in hand a penknife and a notebook. This individual was no different. He placed the root back on the wooden table and with a few short strokes dissected it into halves, then quarters, and finally eighths. At each slice of the knife, he jotted down learned remarks into his pocket notebook.

Suddenly, the owner came storming into the conservatory, arms in a flap, face strawberry red, eyes protruding as if beset by Graves' disease. He was shouting something in Dutch, "Einde, gelieve op te houden, einde!" At first the Englishman could not quite make out what he was saying, for the English even to this day have never had much interest in foreign languages. At last, the Dutchman stuttered in English, "What are you doing!"

"My dear fellow," replied the Englishman, "I am dissecting a most extraordinary onion, pray tell me from where it came?"

"Onion!" cried the Dutchman in despair. "It is an Admiral Van der Eyck!"

"Ah yes, of course!" agreed the Englishman, scribbling the words in his notebook. A few hours later, and to his utter consternation, the Englishman found himself before the local magistrate where he learned to his horror that the onion was no onion at all but a tulip, and worse, it was worth four thousand florins! It was many weeks before the money arrived from England to pay the magistrate's fine. And the gentleman spent many a long night lodged in hellish conditions in the town penitentiary.

To the optimistic Dutch, it seemed as if there had been a paradigm shift, for there had arisen before their very eyes a new source of wealth. Variegated tulips were the vegetative "gold" upon which their economic empire would be founded. Prosperity and Dutch dominance of the world economy were assured. Nobles and paupers all could sell what little they had and invest in tulips. Immense profits were a certainty, for there was barely enough supply to satisfy domestic demand. What would happen when the English, French, and Russians demanded the beauty of these variegated blossoms?

And so the idea *that the passion for tulips would last forever, and that the wealthy from every part of the world would send to Holland and pay whatever prices were asked of them* became firmly lodged in the Dutch psyche. So complete was this belief that jewels, furniture, and land were bartered to obtain tulip bulbs. These bulbs were sold on at a higher price to other speculators who sold them on again. And so the process repeated itself with prices rising ever higher at each trade. By 1635, the price increase was such that *it became necessary to sell them by their weight in perits, a weight less than a grain.* At the zenith of the mania, 12 acres of land around the medieval city of Harlaem were offered for a single tulip root! It was thus that consumed by greed and blinded by unreason, tulipomania gripped the entire nation of Holland.

It came as quite a shock to the pauper and the nobility alike, when in February 1637, prices for the bloom began to fall.

It was not merely that prices fell, for rising and falling prices are part of the natural ebb and flow of supply and demand. No, it was the speed and sever-ity of the decline that mortified the entire Dutch population. For the extent of the tulip delusion had grown so widespread, become so deeply rooted in Dutch society that *nobles, citizens, farmers, mechanics, seamen, footmen, maid-servants, even chimney-sweeps, and old clotheswomen dabbled in tulips.* What once commanded 12 acres of land, overnight, became worthless. As the tulip folly became fully exposed, buying interest evaporated and prices plummeted some 95% in a few months.

The ensuring chaos was ruinous as captured by Mackay:

> Confidence was destroyed, and a universal panic seized upon the deal-ers. Defaulters were announced day after day in all the towns of Holland. Hundreds who, a few months previously, had begun to doubt that there was such a thing as poverty in the land suddenly found themselves the possessors of a few bulbs, which nobody would buy, even though they offered them at one quarter of the sums they had paid for them. The cry of distress resounded everywhere, and each man accused his neighbor. The few who had contrived to enrich themselves hid their wealth from the knowledge of their fellow-citizens, and invested in the English or other funds. Many, who, for a brief season, had emerged from the humbler walks of life, were cast back into their original obscurity. Substantial mer-chants were reduced almost to beggary, and many a representative of a noble line saw the fortunes of his house ruined beyond redemption. (p. 69)

The individual tragedies of wealth evaporated; penury and want were sub-sumed by the longer-term consequences of the tulip boom and bust. For the entire country was thrown into a prolonged deep depression precipitating the end of the Dutch golden age, and with it Netherlanders dream of an economic empire.

A Word to the Wise—You Cannot Rely on the Flynn Effect

It may seem to you that tulip madness was quite clearly irrational, a momen-tary slip in the collective psyche of a normally prudent people. Furthermore, you may even imagine that you could never be caught up in such inanity. Such thoughts are dangerous and should be cast out of your mind. They have led very many intelligent people to make substantial, very painful, life-altering mistakes. This book will help keep you from being one of them.

The Intelligence Quotient (IQ) is supposed to measure intellectual capacity. If it does, then intelligence has been rising steadily since the beginning of the 20th century. This phenomenon is known as the Flynn Effect after James Flynn, the political scientist who first noticed it. The increase in average IQ test scores over generations has been so steep and prolonged that IQ tests have to be made harder every 15 years or so to keep the average individual's score for an age group at 100. Between the years 1952 to 1982, there was a 21-point increase in the IQ scores of American children. If an average individual from 1900 were to take the test today, they would score around 70— that is, they would be classified by the American Association on Intellectual and Developmental Disabilities as having mental retardation.

That the average intelligence of the world is rising, if true, is to be celebrated. That our increasing cerebral activity alone can keep us from the clutches of greed and unreason is unlikely. Take for example, the extraordinary circumstance of Stephen Greenspan, clinical professor of psychiatry at the University of Colorado and emeritus professor of educational psychology at the University of Connecticut. Greenspan earned a clutch of academic degrees including a doctorate in developmental psychology from the famed University of Rochester, a postdoctoral certificate in development disabilities from the University of California's world famous Neuropsychiatric Institute, a master of arts degree from Northwestern University, and a bachelor's degree from Johns Hopkins University. A licensed psychologist in two states, Nebraska and Tennessee, he has also served as a soldier in the United States Army. Given Greenspan's accomplishments, one could quite reasonably expect his cerebral activity to be considerably above average.

During the summer and autumn of 2008, Greenspan was in the painstaking process of putting the finishing touches on his great work *Annals of Gullibility: Why We Get Duped and How to Avoid It*. The book, based on detailed research and his many years of professional and academic experience, was due to be published in early 2009. It contained chapters on gullibility in folktales, religion, politics, criminal justice, science, and most interestingly in finance. To the true believers, within the discipline itself, the ability of the author to understand and explain the world had never been greater.

As Greenspan explained on his personal Web site[5]:

> Gullibility, namely being duped or manipulated by one or more other people, is a very common form of social incompetence, and one that can have very serious consequences for the victim ... This book ... explores ways in which overly-trusting people (or puppets, in the case of Pinocchio) have been duped. My hope is that these stories will contribute

to an understanding of a puzzling phenomenon, namely, why people, sometimes of high intelligence and education, are duped.

With the publication of this new volume, Greenspan would firmly establish himself as the world's leading authority on the developing discipline of gullibility.

On March 12, 2009, in a packed lower Manhattan courthouse, Bernard Lawrence Madoff, the creator of an investment product in which Greenspan had invested a substantial amount of his personal wealth, arose to give his guilty plea allocution.

> Your Honor, for many years up until my arrest on December 11, 2008, I operated a Ponzi scheme through the investment advisory side of my business, Bernard L. Madoff Securities LLC, which was located here in Manhattan, New York at 885 Third Avenue. ... The essence of my scheme was that I represented to clients and prospective clients who wished to open investment advisory and individual trading accounts with me that I would invest their money in shares of common stock, options and other securities of large well-known corporations, and upon request, would return to them their profits and principal. Those representations were false because for many years up and until I was arrested on December 11, 2008, I never invested those funds in the securities, as I had promised. Instead, those funds were deposited in a bank account at Chase Manhattan Bank. When clients wished to receive the profits they believed they had earned with me or to redeem their principal, I used the money in the Chase Manhattan bank account that belonged to them or other clients to pay the requested funds.[6]

At the very time when the Flynn effect had reached an all time high in the United States of America, and on the eve of the publication of Greenspan's great work, numerous investors, gifted, hard working, self-made millionaires, sophisticated hedge funds, established charities, and Greenspan himself, found their wealth evaporated by an investment swindle. And at that, a timeworn Ponzi scheme, first brought to the public's attention over 150 years earlier by Charles Dickens' 1857 novel *Little Dorrit* and made notorious by the Italian emigrant Charles Ponzi in New England during 1920.

Every period of financial turmoil throws up its signature crook. During the economic crisis of 2008, it was Madoff.[7] That he was a thief, serial liar, scoundrel, and conniving dastard is unquestionable true.[8] That he defrauded investors through a Ponzi scheme, as many financial swindlers have done before him, is disappointing, although hardly shocking. That the world's leading authority on gullibility, with his newly published treatise, should be one of Madoff's dupes beggars belief. Of all people, an expert on gullibility should know that if something sounds too good to be true then it is not. Yet, apparently, Greenspan did not.

We bring up the unfortunate affairs of Greenspan to make the point that Bernard Madoff did what swindlers the world over only dream about, he persuaded wealthy and intelligent individuals to hand over their riches with less due diligence than they would carry out before purchasing a bottle of wine at a swanky Manhattan restaurant. So powerful an emotion is greed, and so gripping is the hold of unreason, that a large number of investors placed their entire wealth into his welcoming arms. They expected steady returns, in the order of one or two percentage points a month. Madoff, dressed in his signature charcoal gray suit, smirking through rimless glasses, would refer to it as his *split-strike conversion strategy*.

The investment strategy was to buy a basket of 35 to 50 common stocks listed within the Standard & Poor's 100 price index. This index contains the hundred largest publically traded companies in America. The basket of stocks was chosen to mimic the entire price index, but with an interesting twist. Madoff would opportunistically time the purchase and sale of this basket of stocks. When he was out of the market, the proceeds were invested in United States Treasury bills, the safest of all investments. He also claimed to supplement those investments with related stock option strategies.

As added bait, in order to induce new and continued investments, he would promise certain select prospective investors annual returns as high as 46%. The combined investments were supposed to generate stable returns and to limit losses. Instead, investors' riches funded Madoff's lavish lifestyle and paid off existing investors who wanted to cash out.

In a bull-market rush to get in on the action, concerns over the consistent strength of investment performance were dismissed by his investors. The fundamental rules of risk management were tossed out of the window! Fifteen years of positive performance with only three or four down months was attributed to his superior investment acumen: an investment acumen no other portfolio manager on planet Earth could match! Madoff's returns were simply too good to be true, but no one wanted to believe that, not even Greenspan. The extent of individual and collective unreason surrounding Madoff's investment performance brings to mind the Duke of Wellington's reply to a stranger who greeted him with the words, "Mr. Brown, I believe."

"Sir," said the Duke, "if you believe that, you will believe anything."

By January 2008, Madoff reported around $61 billion under management. But this was the year the market turned sour, very sour indeed. The subprime mortgage crisis of 2008 brought down the great American investment banks. Bear Stearns collapsed, Lehman Brothers Holdings Inc. went bankrupt, Merrill Lynch was sold to Bank of America Corporation, and Morgan Stanley and Goldman Sachs Group Inc. became traditional bank holding companies. With this, and the many other financial calamities of that year, the market was down around 40% by November. Yet, Madoff reported a fictional positive return of 5.6% and imaginary assets under management of approximately $68 billion. Despite these impressive statistics, a large number of his investors wanted their money out; around $7 billion in redemption requests were received during the first week of December alone.

But Madoff's pot was almost empty, only $300 million left. The game was up. Redemption requests could not be met and the whole investment facade tumbled down.

Investors in the Madoff Ponzi scheme, including Greenspan, found themselves cast into misery and very many into penury and want. On the 15th of June 2009, New York City prosecutors filed a collection of impact statements by his victims.[6] It was only then the full extent of the misery wrought became clear to all. One victim wrote:

> According to Madoff's last statement for November 2008, I had $2,300,000 in my family account, $1,200,000 of which was mine personally. Two weeks later, I was bankrupt.

And another:

> My wife and I have lost every dollar of our life savings in Madoff's fraud scheme with no hope of recovery. We have had to sell every asset that we own in order to survive and we don't know how long those proceeds will last.

And yet another:

> We began investing with Madoff in 1993. ... We are now told that he never made any trades at all and that he took every dime we sent him with the express purpose of stealing it.

On and on, Madoff's victims expressed their loss:

> This is not an easy letter to write. I am opening up my families financial status to anyone who wants to see it, which is incredibly humbling and humiliating after years of hard work and major philanthropy. My family's name can be seen on building for the Albert Einstein College of Medicine, The Hebrew Home for the Aged in New Rochelle, and the Hebrew University in Jerusalem. We are benefactors of Lincoln Center and founders the Simon Wiesenthal Center in Los Angeles and too many more charities to mention. Bernard Madoff has robbed three generations of my family. Mr. Madoff seems to have done all he could to protect his family while now I have lost almost everything I have to protect mine.

> He has taken not only my 25 years of savings, but also the lifetime of savings of my 80 year old parents. Keep in mind how he bathed himself and his family in luxury with our money. He ruined lives. He deserves no mercy.

Many members of my family suffered catastrophic losses including those of my 94 year old mother who has been rendered penniless and my legally blind sister most of whose life saving have been wiped out.

It pains me so much to remember my husband, a fine physician, getting up in the middle of the night and going into hospital, in snow and ice and rain, to save someone's life so that Bernie Madoff could buy his wife a Cartier watch.

We have nothing. Only living off social security. I told my father (89) he could not die because I didn't have enough money to bury him.

I am 86 years old, I have a broken knee, I have lung cancer and thanks to Madoff, I am now bankrupt.

In all, there are one hundred and forty one pages of wretchedness and despair. Yet, these pages represented the voices of only a tiny fraction of his victims. The final list was immense, food banks, homeless shelters, homes for the aged, religious organizations, sports clubs for the disabled, services to assist military veterans, and the list went on and on. His swindle spanned the entire globe with in excess of 13,500 different accounts, and losses counted in the billions of dollars. Reflecting the magnitude of the losses, the presiding judge ordered Madoff to forfeit $170 billion in illegally obtained assets.[9] The almost forgotten words of Charles Mackay on tulipmania resonated hauntingly across the centuries to those touched by the ruinous hand of Bernard Lawrence Madoff:

Many, who, for a brief season, had emerged from the humbler walks of life, were cast back into their original obscurity. Substantial merchants were reduced almost to beggary, and many a representative of a noble line saw the fortunes of his house ruined beyond redemption.

That intelligence is sufficient to keep you from succumbing to the charms of greed and unreason is a common misperception—it will not. A word to the wise, you cannot rely on the Flynn effect.

The Unintended Consequences of the Glad Game

Pollyanna Whittier is the heroine of Eleanor H. Porter's 1913 classic children's book series *Pollyanna*. Orphaned as a child, Pollyanna finds herself living with her stern Aunt Polly in the gloomy town of Beldingsville, Vermont. Despite her orphaned status, Pollyanna has a remarkably optimistic attitude, having been taught to play the "Glad Game" by her deceased father. The game consists of looking for and finding the silver lining in all situations.

In one scene from the book, set at Christmas, Pollyanna, hoping for a doll, instead receives a pair of crutches! Making the most of things, she rejoices that she can walk and does not need them. As she explained to Nancy, the housemaid:

> "You see I'd wanted a doll, and father had written them so; but when the barrel came the lady wrote that there hadn't any dolls come in, but the little crutches had. So she sent 'em along as they might come in handy for some child, sometime. And that's when we began it."
>
> "Well, I must say I can't see any game about that," declared Nancy, almost irritably.
>
> "Oh, yes; the game was to just find something about everything to be glad about—no matter what 'twas," rejoined Pollyanna, earnestly. "And we began right then—on the crutches."
>
> "Well, goodness me! I can't see anythin' ter be glad about—getting' a pair of crutches when you wanted a doll!"
>
> Pollyanna clapped her hands.
>
> "There is—there is," she crowed. "But I couldn't see it, either, Nancy, at first," she added, with quick honesty. "Father had to tell it to me."
>
> "Well, then, suppose YOU tell ME," almost snapped Nancy.
>
> "Goosey! Why, just be glad because you don't—NEED—'EM!" exulted Pollyanna, triumphantly. "You see it's just as easy—when you know how!"
>
> "Well, of all the queer doin's!" breathed Nancy, regarding Pollyanna with almost fearful eyes.

Eleanor H. Porter, with Pollyanna, as all great writers do, touches upon the raw essence of humanity; in this case, naive optimism. The Glad Game provides a psychological tool by which a small child, bereft of parents, copes with and ultimately overcomes a hostile world. Playing the Glad Game seems to be an important characteristic of human thought in adults also. It appears to arise because we favor pleasant outcomes over unpleasant ones. As the Scottish political economist Adam Smith tersely observed:

> The over-weening conceit which the greater part of men have of their own abilities, is an ancient evil remarked by the philosophers and moralists of all ages. Their absurd presumption in their own good fortune, has been less taken notice of. It is, however, if possible still more universal. There is no man living who, when in tolerable health and spirits, has not some share of it. The chance of gain is by every man more or less over-valued, and the chance of loss is by most men under-valued, and scarce by any man, who is in tolerable health and spirits, valued more than it is worth.

Psychologists have long observed that people generally give themselves a much higher chance of surviving natural disasters than those around them.

When it comes to ill health, heavy smokers view other smokers as more likely to suffer from the well-known consequences of smoking than they are themselves. Students consistently overestimate their ability to perform in examinations. Indeed, in many cases the correlation between self-rating of ability and actual performance is meager. Individuals have a tendency to overrate themselves and the likelihood of achieving favorable outcomes. It is an inherent human characteristic. Not only is such optimism pervasive, it may not be evident to those who exhibit it.

Take for example the case of McArthur Wheeler[10]:

> In 1995, McArthur Wheeler walked into two Pittsburgh banks and robbed them in broad daylight, with no visible attempt at disguise. He was arrested later that night, less than an hour after videotapes of him taken from surveillance cameras were broadcast on the 11 o'clock news. When police later showed him the surveillance tapes, Mr. Wheeler stared in incredulity. "But I wore the juice," he mumbled. Apparently, Mr. Wheeler was under the impression that rubbing one's face with lemon juice rendered it invisible to videotape cameras.

Regrettably, Wheeler's behavior is not that unusual. There are many individuals who are highly unskilled and highly unknowledgeable, but do not necessarily realize it. Instead, they hold overly favorable views of their abilities. They suffer from the double curse of incompetence, lacking both the knowledge and expertise necessary to accurately assess their ability.

Insight was thrown on this seemingly bizarre phenomenon by two American psychologists, Joseph Luft and Harry Ingham. During the early to mid-1950s, Ingham and Luft developed a cognitive psychological tool known as the Johari window.[11] The window consists of four areas (see Figure 1.1).

1. *Open* region of what is known by a person and others about him or herself.
2. *Hidden* region of what is known by a person about him or herself which others do not know.
3. *Blind* region of what is unknown by a person about him or herself which others know.
4. *Unknown* region of what is unknown by a person about him or herself and unknown by others.

Entire organizations can be subject to this bias too. *Harper's New Monthly Magazine* reflecting in 1876 on the tulip mania declared:

> Such delusions are most fertile in an age of financial ignorance. There has been too large a development of educated common-sense, too much of a study of the principles that underlie the making of money, and, above all, the press is too enlightened and powerful to permit them to beggar

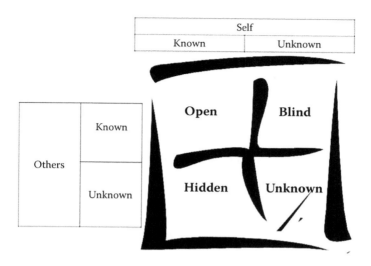

FIGURE 1.1
The Johari window. (Luft, J. and Ingham, H., 1955, The Johari Window, a Graphic Model of Interpersonal Awareness, *Proceedings of the Western Training Laboratory in Group Development,* Los Angeles: UCLA.)

whole nations as they once did. The financial crises of the present day are short-lived and confined to commercial centres, but three centuries ago they ruined whole peoples.

Flawed self-assessment can as Adam Smith reflected lead to *the contempt of risk and the presumptuous hope of success.* Few tulip investors suspected the boom would have such a catastrophic end, or that the Dutch claim to economic greatness would end miserably some years later with the lynching of the prime minister, Johan de Witt, by an angry mob of de Witt's own countrymen. Rather unfortunately, investors fueled by the presumptuous hope of success continuously take actions that are contemptuous of risk. Indeed, entire societies can find themselves playing the Glad Game. And this is not limited, as *Harper's New Monthly Magazine* presumed, to the distant past.

In the United States of America during the late 1990s through the mid-2000s, the Clinton and then the Bush administration encouraged home ownership by urging banks to relax lending standards. Subprime mortgage borrowers, many barely eking out a living, bet their futures on a housing bubble they believed would never burst. Around the same time, banks adopted new loan safety guidelines. These guidelines called BASEL II after the Swiss city in which they were negotiated, deemed owner occupied mortgages an ultra-safe investment. Banks that brought mortgages as opposed to say corporate debt could use much more of their own money, maintain lower capital reserves, and therefore earn higher returns. Investors too placed their faith in mortgage backed securities, bundled packages of mortgages. Banks and financial institutions built entire business lines devoted to extending loans to individuals who had very little means. So easy was the lending that at one point it seemed possible to obtain a mortgage without any evidence whatsoever of income of any kind. The joke among mortgage brokers was as long as you had breath in your body you were "approved."[12] Although a pulse was not strictly necessary:

> That pulse thing, that was also optional. Like this case in Ohio, where 23 dead people got mortgages. And even some of the living weren't exactly what you'd call a good credit risk. The name of this new loan, remember, was No Income No Asset. People in the industry called it a liar's loan. They expected people to lie.[13]

The tax code incentivized speculation by allowing sellers to pocket up to $500,000 in capital gains tax-free and repeat the feat every 2 years ad infinitum. Fanned by a gaggle of reality television shows, radio infomercials, and newspaper real estate "rags to riches" articles, the entire United States of America found itself playing the real estate Glad Game. The television shows would detail the day-to-day trials of a "regular Joe" investor in residential real estate. A property would be purchased, refurbished, and then sold on at immense profit. In many cases, the whole cycle of buy—refurbish—sell took a matter of weeks. Numerous books were published on how to cash in; and all manner of people from all walks of life were so convinced of the profitability of the opportunity that they rushed to buy, refurbish, and sell properties. Known as *flippers*, many appeared to have no means whatsoever. But this did not matter; they could be serviced by *no money down* mortgages.

People who said the prices could not possibly go higher watched with chagrin as their friends and relatives made enormous profits. The temptation to join them was hard to resist and real estate investment clubs flourished. However, as all bubbles do, the housing bubble eventually burst, and in doing so, brought about widespread panic and financial distress not seen since the great stock market crash of 1929. An overdemand and oversupply of bad housing loans ultimately lay at the root of the crisis. During the boom years, Americans were seduced by the illusion of easy money, and a blind faith that house prices could only go up. After the collapse, as was the case with tulip madness in Holland, it seems obvious. It was always too good to be true. A number of analogies spring to mind, perhaps none more boldly than the Yorkshire phrase, *There's none so blind as them as can't see*. We understand now why Nancy, the maid servant, looked at Pollyanna with "fearful eyes" for it is possible to play the Glad Game and not be aware that one is playing it.

Unreason in the form of overly positive self-evaluations, exaggerated perceptions of control or mastery, and unrealistic optimism is an inherent weakness of human nature. It can lead intelligent and educated people to make investment decisions that ignore the fundamental rules of risk management with ruinous effect. The sheer strength of greed and his even eviler twin brother unreason's grip, once it takes root, brings to mind the White Queen in Lewis Carroll's *Through the Looking Glass*:

> "Why sometimes," said the White Queen to Alice, "I've believed six impossible things before breakfast."

Individual investors playing the Glad Game have believed many more impossible things than that. In an ironic twist, on the death of Pollyanna's creator, Eleanor H. Porter, in May 1920, Charles Ponzi's great swindle was just gathering steam. The people of New England had embarked on a new Glad Game.

But You Have to Remember Ivar Kreuger of Kalmar!

Not long after I had written an amazing article outlining a new innovative approach to determine when to buy and sell gold (see www.NigelDLewis.com), I received a phone call from a friend who had recently moved to a teaching position in New York City. He was excited about an incredible piece of news. Something fascinating had just happened, and he thought that as an author on investing I might be able to enjoy his good fortune.

My friend, let's call him Bernie, had, quite by chance, happened upon an article in a free local newspaper. The article was about something "much better than gold"—potash! Turns out the price of potash had recently reached an all time high. Bernie knew of my research in the area of diversified investing and was excited to tell me about his discovery. He sincerely wanted to know my view on the merits of investing half of his wealth in potash.

This was mildly surprising to me because Bernie was a very conservative investor. His entire retirement account was in government and corporate bonds, and this even though he was still many years from retirement. Potash seemed inappropriate given his risk appetite. And I told him so.

Eventually, unable to get me excited about allocating a significant proportion of his wealth to potash, Bernie admitted something I regarded as rather foolish. He had attended, that very evening, a free seminar. It was at a local discount motel whose address was specified in the advertisement. Packed in with two or three hundred other eager investors, he was still buzzing with enthusiasm. The evening had begun with the host, a sharp-faced man, eyes black as coal, in a deep blue pinstriped suit, strutting backwards and forwards on a makeshift raised stage. Flapping his arms about excitedly, as Icarus must have done before his fatal plunge, the suited gentleman extolled the virtues of investing in "virtual potash." Apparently, it was easy, risk-free, and best of all, profits were guaranteed even if the price of *real* potash fell. But there was a catch; only twenty people from the audience would be selected to participate. The purpose of the meeting was to determine who the lucky selectees would be.

The air was electric with excitement. A free lunch was about to be served and my friend was very hopeful he would leave the meeting well fed. At seemingly random points in time the host would ask for members of the audience who had used his virtual potash system to come onto the stage. And sure enough, somebody would step forward and recount an amazing story of rags to riches. Bernie was particularly taken by a Hispanic lady. She was tall and slender with hazel brown, darting eyes, little more than thirty-five-years-old, and barely able to speak English. She had not graduated high school, and as a consequence had spent her entire career in menial labor. Her very last job was gutting chickens at an unregistered meat packing plant just outside Des Moines, Iowa. Through an event such as this one, she had discovered the virtual potash system. Now a millionaire, work was off her agenda for life!

Curious to find out more details, I asked, "But, Bernie, how did she do it, what is the secret sauce?"

My friend replied, "Buy potash but don't take delivery! The company will store it for you," he whispered cautiously. Then, as if somebody might over-hear our conversation, he murmured almost inaudibly, "This allows them to guarantee an additional 20% return if the price goes up and your money back if the price goes down. But you need to lock the investment in for 18 months. I invested $75,000. What do you think?"

Seventy-five thousand dollars represented virtually his entire savings! Bernie, a divorced man with three kids, was not a young teacher, and New York City is a very expensive town. I thought I knew what had happened but told him that, if I were to explain things properly, he would have to listen to a story of mine involving a Swedish individual by the name of Ivar Kreuger.

"Ivar who?" he asked quizzically.

"But you have to remember Ivar Kreuger of Kalmar," I said.

Apparently, he did not.

There are few areas of life where skepticism is more important than how to invest. Yet, very few financial advisors, investment professionals, or private investors remember Ivar Kreuger of Kalmar or the lessons his saga teach, and this is a shame. Kreuger was one of the wealthiest men in the world in his day.

He was born in the Baltic Sea town of Kalmar in the southeast of Sweden on March 2, 1880. The medieval town protected by the grand castle of King Magnus Ladulås is the seat of the Kalmar Diocese with its great cathedral as the centerpiece. Kreuger would have passed this magnificent classicistic building many times as he grew up, perhaps even entering on occasion to kneel in prayer before the altar. But its message of hope, redemption, and love found in Kreuger a barren, stony heart. Today, and in a strange way echoing the anguish and misery surrounding his memory, it lies empty, the only cathedral in the entire country that is without a bishop.

The Kreuger family ancestry, which contains some of Sweden's great merchants, can be traced back to the German baker Johan Kröger, who immigrated to Kalmar in 1710. Ivar's father, Ernst August Kreuger, was a Russian consul. Kreuger graduated at the tender age of 20 from the Teknisk Högskola in Stockholm with a dual master's degree in mechanical and civil engineering. He was intelligent, ambitious, and extremely charming. His family operated a number of match manufacturing factories scattered around the town of Kalmar—The Fredriksdahl Match Manufacturing Company, Mönsterås Match Manufacturing Company, and Kalmar Match Manufacturing Company.

Matches were a staple product, held in similar regard to bread and milk. World demand was in excess of 20 billion boxes per year. They were used for lighting cigarettes, gas lamps, fireplaces, and stoves.

When Kreuger joined the family match making business, he floated the idea of turning their company into a stock corporation in order to raise capital for expansion. This was the game-changing idea upon which Kreuger's empire would be built. Through a series of deft acquisitions and innovative manufacturing and design ideas, he transformed the family business into the world's largest producer of safety matches. By the end of 1930, his companies made 90% of the world's matches, controlling 250 factories in 43 countries in Europe, North and South America, and Asia; hence his sobriquet, the *Match King*. It was an astounding feat achieved in large part by his move into the arena of international finance.

On June 28, 1914, Archduke Franz Ferdinand of Austria, heir to the Austro-Hungarian throne, was assassinated precipitating the start of the First World War. By its close, on the eleventh hour of the eleventh day of the eleventh month of 1918, the imperial powers of Germany, Russia, Austria-Hungary, and the Ottoman Empire—had been defeated. The economies of the remaining great powers of Europe were in disarray. Among the devastation, Kreuger spotted an opportunity. Europe was in need of capital and the United States had a surplus. The question was how could he benefit from this.

His *great idea* was to offer European governments ultra cheap loans in exchange for the right to set up match manufacturing monopolies in their respective countries. The source of the loans would be America's surplus capital. With this idea in mind, during the autumn of 1922 he set sail aboard the luxury Cunard liner *RMS Berengaria* for the United States of America. On his arrival in New York City, he could sense a mood of euphoria beginning to grip Wall Street. It was the decade of jazz and bathtub gin, the $5 work day, the Model T Ford, the first transatlantic flight, the movie, and Calvin Coolidge's declaration *America's business was business*. The Roaring Twenties were underway. America had entered its modern era. Krueger, it appears, had arrived at precisely the right time. His intuition had been correct. Americans were ready to invest in Europe.

Had he arrived a decade earlier this would have not been the case. Before World War I, only a tiny fraction of Americans invested in the stock exchange. Wall Street was denounced by populist politicians as the devious concoction of the *robber barons*, Vanderbilt, Gould, Dew, and the House of Morgan, all of whom had amassed huge personal fortunes. Their wealth was perceived by many to have been gathered as a consequence of anticompetitive, questionable, or outright unfair business practices. Wall Street was thus viewed by the general populous with a mixture of loathing and fear.

In order to support her war efforts the United States issued a series of *liberty bonds*. The first issue raised $2 billion with over 4 million subscribers. The second raised $9 billion with 9.4 million people subscribing. Liberty bonds rapidly became a symbol of patriotism. By the conclusion of the war in 1918, Americans' view about Wall Street had radically changed. Having been buyers of liberty bonds, Americans had lost their fear of investing. Stockbrokers began to open offices not on Wall Street but on main street. Soon stock market trading became America's favorite pastime. News of stock market millionaires served only to fuel the investment frenzy. In almost every year from 1924 through 1929, the upward momentum in the Dow Jones industrial average seemed unstoppable.

In America, Kreuger met with the investment bankers Donald Durant, Federic W. Allen, and Jerome Davis Greene of Boston-based Lee, Higginson & Co. The investment bank had grown to national importance having played a prominent role in financing the development of U.S. railways and the 1910 reorganization of General Motors. By the 1920s, its salesmen covered the entire nation. The strategy was to raise capital by issuing debenture bonds to the American public through an American company, Kreuger & Toll International Match Corporation.

The bonds offered incredible rates of return, as high as 25% per annum. However, before the issue of millions of Kreuger securities to investors could be authorized, an audit would have to be conducted. Kreuger, ever resourceful, had three Swedish audit reports prepared in advance. So charmingly persuasive was he and so great an economic opportunity did this deal represent, that Lee, Higginson & Co. waved normal audit practice. They relied on the Swedish audit reports. Common practice was to use a British or American firm to conduct audits. The investment banker Jerome Davis Greene was so taken by the profitability of the opportunity and the certainty of immense profit that he invested a substantial portion of his personal fortune in the "Kreuger paper." Other investors also took the bait, and Kreuger raised hundreds of millions of dollars. Loans were extended to Poland, France, Spain, Germany, and Turkey, among others.

The principal difficulty for Kreuger was these loans only paid around 6% while profits from his expanding match making monopolies were not as strong as expected. He was unable to cover the payouts to his American bondholders. The only way to make up the gap was to raise more cash or speculate, and he did both. As with the many swindlers before and after him, Kreuger had an utter disregard for accounting niceties. He personally prepared the financial statements of his companies, and he did so without any reference whatsoever to ledgers, sales, or other financial facts one might naturally expect would be required in order to prepare accurate statements. His Swedish accountants would then dutifully prepare the books to match these *Kreuger inspired* financial statements. To raise capital, he attracted investors by offering huge dividend payments. These were necessary to ensure the continued sale of new securities. The continued sale of new securities was necessary to make the dividend payments. It was a never-ending cycle of deception. The whole system depended on a constant input of new capital. It was essentially a giant pyramid scheme.

So attractive were the returns of Kreuger paper that they became the most widely held securities in the United States. To cover his tracks, Kreuger constructed increasingly complex transactions between the various entities of his empire that no one else except he knew about. His statements carried many intangible assets such as monopoly rights in various countries. He created fictional companies, which on paper were extremely profitable. In all, around 250 subsidiaries and trading concerns were created. When the coffers were empty, he forged Italian bonds to create the illusion of wealth. Herr Kreuger and Herr Kreuger alone was the only person who had a clear understanding of how the complex tangle of companies and deceit were tied together.

The great stock market crash of 1929 and the ensuing depression created a liquidity squeeze. Kruger's great swindle, as did Bernie Madoff's some 80 years later, became exposed as the whole investment facade slowly began to crumble. The end finally came on the morning of March 12, 1932. Krueger had taken up residence at No. 5 Avenue Victor-Emmanuel III, Paris. His creditors planned to confront him that very day. The game was up; Kreuger knew it. But he could not face his foes. He penned a short note then shuttered the blinds. The consequences of lying, cheating, and stealing his way around the globe must have weighed heavy on his mind. It had been an unbearable few years, simply intolerable. He lay down upon his bed, drew aside his waist jacket, and with a small pistol, a Browning 9-mm, shot himself straight through the heart. He died shortly thereafter.

Kreuger's financial empire was an enormous Ponzi scheme. Within a month of his death, most of his business empire collapsed into bankruptcy as his complicated web of forgeries, theft, fraudulent bookkeeping, and fictitious companies was uncovered. His houses, boats, furniture, paintings, and remaining assets were sold off at auction. Nothing was left behind for his family except his contemptible reputation for treachery and thievery. In the United States, his debentures proved almost worthless and news of his demise ultimately led to the collapse of the investment bank Lee, Higginson & Co. Jerome Davis Greene, the investment banker, was cast out into unemployment virtually penniless. Greene was fortunate, for through a friend, he was able to secure a professorship in international relations at Aberystwyth, Wales. Others were less fortunate. Kreuger had swindled the public directly out of at least $560 million.[14] His victims included thousands of investors, university endowment funds, and banks. In a way, the series of despicable acts carried out by Ivar Kreuger substantiates Warren Buffet's famous aphorism:

It's only when the tide goes out that you see who has been swimming naked.

Ivar Kreuger was a gigantic figure of his era, a titan of his time. His face graced the cover of newspapers and magazines. He dined with presidents and kings, was knighted by France, served as peacemaker for the League of Nations at The Hague, was a frequent visitor to President Hoover's White House, and was suggested as a candidate for the Nobel Peace Prize. One of the most incredible things about this story is that he has been totally forgotten by modern day investors and investment professionals; his name and despicable deeds washed away by the rivers of time. It seems the memory of financial swindlers fades rapidly across the generations.

My teacher friend had invested in a Ponzi scheme, much like the hapless investors in Kreuger paper, but this time the foil for thievery was virtual potash. But still giddy from his recent encounter at the seminar, he was not ready to admit this. Another illustration was called for, this time even more shocking.

The Baptist Foundation of Arizona (BFA)[15] was founded in 1948 with the support of Arizona's largest Baptist denomination, the Southern Baptist Convention. As a nonprofit, tax-exempt, charitable corporation the initial focus was on fund raising to support Baptist causes in Arizona. Initial donations were used to *plant* new churches and offer specialized ministerial services especially to children and the elderly. By the late 1950s, the foundation's benevolent activities had expanded considerably and there was a growing need to appoint a full-time president. In 1962, a young preacher by the name of Glen Crotts was chosen to take the helm. He remained as president for 20 years, only stepping down in 1982 to make way for his son, William Pierre Crotts.

For many years, W.P. Crotts lay in the shadow of his father. Pastor Glen Crotts, first president, organizer of the respected foundation, planter of churches, and provider to needy Baptists, had become almost a mythical figure. W.P. Crotts, an attorney, albeit of questionable ability, had been surrounded his entire life by the prestige and privilege that comes with being the son of a high-flying Baptist pastor. Perhaps he had become accustomed to it, maybe he even thought it was his birthright. Either way, on his succession, he gathered around himself a clique of trusted advisors; Tom Grabinski as legal counsel and Donald Deardoff as controller. They quickly set about developing a new strategic vision for the foundation.

The vision would totally discard the foundation's benevolent roots, replacing it with aggressive speculation and capital accumulation. This was a surprising strategy given that Southern Baptists have a long tradition of abstinence from both alcohol and gambling. Yet, W.P. Crotts, Grabinski, and Deardoff devised a plan to speculate on Arizona real estate. W.P. Crotts, in common with all businessmen of big ideas, was eager to raise as much capital as possible, perhaps to create the impression that he was a moneymaker, worthy of being entrusted with the presidency once held by his father. The question was how to raise enough capital to make speculation worthwhile?

Arizona, one of the *sunbelt states*, grew in population by over 100% between 1945 and 1960. It grew by around 40% over the 10-year period 1990 to 2000. As a consequence, there had been explosive growth in retirement communities. Among this group were many thousands of retired Baptists whose retirement accounts were brimming with a lifetime of savings. This observation did not escape the notice of W.P. Crotts, Grabinski, and Deardoff. In 1984, their plan to raise capital began to take shape. The Baptist Foundation offered a *tremendous opportunity to be found faithful* to those who invested in their low-risk promissory notes. Indeed, one of their marketing leaflets stated:

> We are a ministry dedicated to serving the Lord and furthering Southern Baptist and other Christian causes. We re-invest your money and the profit we earn goes to further such ministries as Christian education, care for children and senior adults, missions and new church starts. Your investment actually touches the lives of countless numbers while you earn a very attractive interest rate.[16]

Investors had little reason to doubt their good intentions. So powerful was the pull of faith-based investing that between 1984 and 1985 alone, assets under management grew from $7.2 million to $211 million. Investors deposited funds into investments with names such as *"Easy Access Investment,"* and *"Maximum Value Performance Note."* By the mid-1990s, around 11,000 individuals had wealth secreted in a variety of the foundation's faith-based investments.

W.P. Crotts, Grabinski, and Deardoff had hit the jackpot. They used the foundation as a bank, borrowing money from investors at high rates of interest, and then relending it for speculative real estate deals that enriched their friends and associates. Investors were led to believe the *profits* would go to Christian causes. In fact, over an entire 50-year span the foundation had returned only around $1.3 million of its own money to the Baptist community, indeed:

> Companies controlled by one sitting BFA director and two former directors have received nearly $140 million worth of loans in complicated real estate and stock transactions with BFA. Public records in several states indicate that for at least 10 years, BFA has served as a seemingly bottomless pool of capital for this cadre of insiders. BFA's managers appear to have gone to great lengths to disguise the insider loans—creating a labyrinth of 63 for-profit and non-profit companies.[17]

Real estate is subject to boom and bust, and this occurs with monotonous regularity. Speculative investments by the foundation in the boom years proved to be disastrous failures when Arizona's real estate bubble burst in 1989. To cover their losses, W.P. Crotts, Grabinski, and Deardoff engaged in an outrageous Ponzi scheme, one whose primary victims would be the elderly and infirm. That this was the case did not seem to matter much to W.P. Crotts, Grabinski, or Deardoff. Indeed, that their victims would be primarily Southern Baptists, the very group that had founded their organization, appeared to have no restraining effect on their actions. Instead, gambling and alcohol forbidden for the faithful were eagerly embraced as they took to the scam like a hungry dog to a bone. The trio enjoyed a lavish "bling bling" Hollywood movie star lifestyle while shamelessly encouraging a steady influx of new investor money. This money was used to pay interest on old money. It was in essence the same detestable swindle entered into by the scoundrel Ivar Kreuger and the glutinous dastard Bernie Madoff. This time, and quite shockingly, the Baptist Foundation of Arizona was fleecing their flock!

A loud cry of amazement went up from the community when it was discovered that W.P. Crotts had been "gambling" on property prices. The shock of disillusionment dealt a heavy blow. For as well as livelihoods, he had destroyed faith, that faith which the Baptist community had placed in the foundation as a glowing example of faith-based investing. The total disregard for all ethical considerations, which made W.P. Crotts, Grabinski, and Deardoff utterly dishonest in their business dealings brings to mind the words of Isaiah:

> Yea, they are greedy dogs which can never have enough, and they are shepherds that cannot understand: they all look to their own way, every one for his gain, from his quarter.
>   Come ye, say they, I will fetch wine, and we will fill ourselves with strong drink; and tomorrow shall be as this day, and much more abundant. (KJV, Isaiah 56:11–12)

Just about the best which may be said of their seamy conduct is that it was no better or worse than the fraud and deceit practiced by any other thief intent on stealing and concealing by use of a Ponzi scheme. Such schemes only collapse when their sources of new money are shut off. Unfortunately, in the case of the Baptist Foundation of Arizona, this was not before the depravity of W.P. Crotts, Grabinski, and Deardoff had defrauded 11,000 investors out of $550 million.[18,19]

I brought up the case of Ivar Kreuger and the Baptist Foundation of Arizona to make the point to my teacher friend that such swindles are often perceived as extremes, unusual rarities, located in the far left hand tail of the bell-shaped curve. Investors, it is argued, should not be concerned about these once in a lifetime events. The fundamental rules of risk management simply do not apply. Alas, this is a misperception. The actions of individuals such as Bernie Madoff, Ivar Kreuger, and W.P. Crotts are neither rare nor special. The investment landscape is literally brimming with scoundrels who operate Ponzi schemes. Neither charitable status nor the goodness of the works will deflect these sorry individuals from their greedy scheming.

As Tamar Frankel, who has studied these matters extensively, concludes[20]:

> The amounts involved in Ponzi schemes are usually very large. They catch in their net billions of dollars from very wealthy as well as less wealthy individuals and institutions. The annual losses from Ponzi schemes in the United States vary. Based on litigated court cases, the year 2002 showed the largest amount of losses—over $9.6 billion. Each of the years 1995 and 1997 showed losses more than $1.6 billion. Each of the years 1996, 1990 and 1976 showed losses of over $1 billion. These numbers, however, represent only those cases that were litigated in the courts, and do not show the losses outside the courts and on the international scene. ... Ponzi schemes are not unique to the United States. They have been highly successful in Romania, India, Albania, Russia and England. Thus, Bernie Madoff's scheme is far from special, although it is quite large (see Figure 1.2).

And what of my teacher friend Bernie? He was so eager to believe in the virtual potash system that he dismissed my warnings as "knee jerk cynicism." Unreason in the form of self-deception had him in its firm grip. Greed left him with the false assurance of a certain killing. Risk management was a bunch of baloney! Bernie sincerely believed he would soon be retired, laughing at my scorn while sipping a cocktail on a Caribbean beach. He lost every

Ponzi schemes are more common than you think!

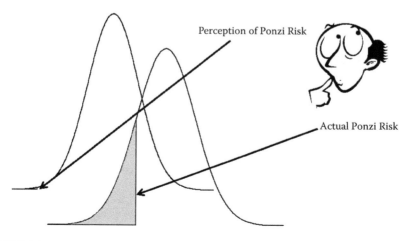

FIGURE 1.2
Distribution of Ponzi risk.

penny! And this is precisely why unreason is the even eviler twin brother of greed.

Endnotes

1. See *Harper's New Monthly Magazine* (1876).
2. See *Spruce Roots Magazine* (http://www.spruceroots.org).
3. See the historical archives of the John Innes Centre (http://www.jic.ac.uk).
4. The results were published in Cayley, D.M. (1928), "Breaking" In Tulips, *Annals of Applied Biology*, 15:529–539. doi: 10.1111/j.1744-7348.1928.tb07775.x. She uncovered two key findings: (1) "Breaking" in tulips is infectious and can be induced by bringing tissue of a normal bulb in contact with tissue from a "broken" bulb during the resting stage; and, (2) The extent of breaking is proportional to the amount of infected tissue introduced.
5. See Stephen Greenspan Ph.D. (http://www.stephen-greenspan.com).
6. The case is *U.S. v. Bernard L. Madoff*, 09-cr-213 (DC) June 12, 2009, U.S. District Court for the Southern District of New York (Manhattan).
7. The size of his scam, around $50 billion, its duration over almost two decades, and his position as former nonexecutive chairman of the NASDAQ stock exchange, struck a deep blow into the heart of the investment management industry. It was the largest investor fraud ever committed by a single person.
8. Bernard Lawrence Madoff engaged in wholesale fraud, repeatedly lied under oath, and filed false documents to conceal his fraud. Starting as early as the 1980s, he engaged in deceit and deception, abused his position of trust, and misled individuals, charities and other institutions. Between December 1995 and December 2008 approximately 1,341 accounts suffered net losses totaling at least $13 billion.
9. The U.S. District Court, Southern District of New York also stipulated that his wife, Ruth Madoff, be stripped of $85 million in assets.
10. Quote taken from Kruger and Dunning (1999).
11. See Luft, J. and Ingham, H. (1955), The Johari Window, a Graphic Model of Interpersonal Awareness, *Proceedings of the Western Training Laboratory in Group Development* (Los Angeles: UCLA).
12. See, for example, the Center for Responsible Lending (2006) who report that "stated income" loans, in which borrowers' income was simply affirmed without supporting evidence, were as much as 26% of some loan pools.
13. *This American Life*, "Global Pool of Money Got Too Hungry," National Public Radio, May 9, 2008.
14. Approximately $8.4 billion in 2009 prices. This is a conservative estimate.
15. This was the longest criminal trial in Arizona history.
16. See *Phoenix New Times* (http://www.phoenixnewtimes.com).
17. The large chain of profit and nonprofit companies was designed to conceal their utter failure as investment professionals and hide traces of their fraudulent manipulations.

18. In August of 2006, after a 10-month trial, a jury found against the following individuals:
    William Crotts (Case Number: CR2006-006183);
    Count 1: Fraudulent Schemes and Artifices: 8 years prison.
    Count 31: Knowingly Conducting an Illegal Enterprise: 7 years prison (to run concurrent with the sentence imposed in Count 1).
    Thomas Grabinski (Case Number: CR2006-006183);
    Count 1: Fraudulent Schemes and Artifices: 6 years prison.
    Count 31: Knowingly Conducting an Illegal Enterprise: 5 years prison (to run concurrent with the sentence imposed in Count 1).
    Lawrence Dwain Hoover (Case Number: CR2006-006183);
    Count 1: (Amended) Fraudulent Schemes and Artifices: 5 Years Probation.
    Restitution: $500,000.00 ordered and paid in full to the Baptist Foundation of Arizona Liquidation Trust.
    Donald Dale Deardoff (Case Number: CR2001-006926);
    Count 1: Fraudulent Schemes and Artifices: 4 years prison.
    Count 2: Fraudulent Schemes and Artifices: 4 years prison.
    Incarceration terms ordered in Counts 1 & 2 shall run concurrent.
    Restitution: $150,000,000.00 ordered to be paid to the BFA Liquidation Trust.
    Edgar Alan Kuhn (Case Number: CR2001-006928);
    Count 1: Facilitation of a Fraudulent Scheme or Artifice: 3 years probation.
    Count 2: Facilitation of a Fraudulent Scheme or Artifice: 3 years probation.
    Count 3: Facilitation of a Fraudulent Scheme or Artifice: 3 years probation.
    The probation terms ordered in Counts 1, 2, and 3 shall run concurrent.
    Restitution paid in full by defendant in 2001.
    Jalma Hunsinger (Case Number: CR2001-006927);
    Count 1: Facilitation of Illegally Conducting an Enterprise: 3 years probation.
    Count 2: Facilitation of Illegally Conducting an Enterprise: 3 years probation.
    Count 3: Facilitation of Illegally Conducting an Enterprise: 3 years probation.
    The probation terms ordered in Counts 1, 2, and 3 shall run concurrent.
    Restitution paid in full by defendant in 2001.
    Richard Lee Rolfes (Case Number: CR2006-006183);
    Count 3: Facilitation of a Fraudulent Scheme or Artifice: 3 years probation.
    Restitution: $25,000.00 paid February 2, 2007.
    Harold Dwayne Friend (Case Number: CR2006-006183);
    Count 31: (Amended) Attempting to Assist in a Criminal Syndicate: 3 years probation.
    Restitution: $240,000.00.
19. On June 10, 2009, the Arizona court of appeals upheld the convictions.
20. Presented before the Committee on Financial Services of the U.S. House of Representatives on January 5, 2009.

2

The Maleficent Hand of the Men in Gray Suits

Unreason Abounds in Places Where It Must Not

To begin to appreciate why this is so, it is helpful to consider a fascinating experiment carried out in the spring of 1937 in Altenburg, Austria. It was conducted by two zoologists, Konrad Lorenz and Niko Tinbergen, early founders of the discipline of ethology—the study of animals in their natural setting. They were fascinated by the idea of innate species-specific behavior patterns—actions inherent to individual species in the same way as tigers have specific claws or sharks have unique teeth. Sometime earlier, while walking along Jesus Lane in Cambridge, they had discussed their theories and conjectures. Now in Austria, on a 3-month spring break, they decided to investigate further.

Between two tall trees, they strung a heavy rope. Attached were an assortment of birdlike shapes cut from cardboard which could be pulled along the rope to mimic the motion of birds in flight. Tinbergen and Lorenz gathered a gaggle of young geese, turkeys, and ducks underneath the rope. The cardboard shapes were pulled at varying speeds backwards and forwards along the rope. The ducks and geese paid little attention. The turkeys were different. One cardboard cutout, pulled in a particular direction, resulted in great agitation and the sounding of their alarm squawk. Tinbergen and Lorenz tried it again, same response, and again. Each time they pulled the cardboard cutout in a particular direction, the turkeys would become agitated and sound their alarm squawk. What could explain the unusual reaction of those young turkeys?

The cardboard dummy had what looked like wings placed near its rear. When pulled in one direction, it resembled a bird with a short neck, symbolic of a predator such as a hawk or eagle. When pulled in the opposite direction, it resembled a long-necked bird such as a goose (see Figure 2.1). If pulled slowly in the direction that resembled a goose in flight the turkeys remained calm. If pulled slowly in the opposite direction it would elicit alarm squawks and agitation among the young turkeys. The researchers concluded the response of the turkeys to the configuration and direction of the shape was an innate response to an environmental cue signaling *predator*. This was truly a remarkable finding.

How to Fool Turkeys

FIGURE 2.1
Lorenz and Tinbergen bird experiment.

We may scoff at the turkey for being fooled by a cardboard cutout. How silly it seems that they will squawk and flap when it is pulled in one direction, but not the other. But ethologists inform us this response is far from unique to turkeys. Herring gulls can be tricked into regurgitating food using a red knitting needle with a white band around the tip. Thrush chicks open their mouth and thrust their head forward for food. The chick that gapes the widest is fed first. The gaping response can be elicited by a light touch. Even the most simple of organisms like bacteria and protozoans demonstrate simple inherited behavior. Bacteria will migrate toward or away from light or salt. Stentor, a very simple type of protozoan, will react to prodding by moving away.

Among investors too, perhaps a more complex version of the same effect holds, and this is particularly relevant when it comes to financial regulators. For regulators, dressed in their uniform of gray, offer up a soothing illusion of order and calm. For the most part, they operate in the shadows, glimpsed only on rare occasions as they scurry about their business. Rarely do investors consider their existence. And this is so even though many invest their wealth in the comfortable notion that the men in gray suits are working diligently on their behalf, will protect them, and will come to their aid if they should cry out. And this is a terrible mistake. For, if the regulator turns out to be *made of cardboard,* the sense of safety will prove a costly fiction; costly because the perception itself may encourage risk which might otherwise be avoided. A cardboard financial regulator whose actions induce risk rather than reduce it may seem, at first blush, absurd. Unfortunately, it is not.

The Conspiratorial Regulator

Nobody could have imagined Darrel W. Dochow would be caught up in a national scandal, not least Dochow himself. That he and his superiors would be pursued by an angry mob of investors and taxpayers, or that the accusatory finger of vote hungry politicians would be raised and pointed in his direction seemed inconceivable.

By all accounts, he was a pleasant enough individual. Photographs of him at work show a mature, rotund man dressed in a dark single-breasted suit, impeccably pressed seashell white shirt with a modest conservative tie. His face vaguely resembled that of an elongated pug, a bulbous nose with sunken dark eyes, which offered only the slightest hint of curiosity. His head, pasty and pink, was partly bald with combed back silver-gray hair, which exposed a dome shaped forehead. His countenance and bearing fitted precisely what one might imagine of a civil servant. Much as dog owners and their animals are said to grow to resemble each other, Dochow, after more than 30 years of government regulatory service, had come to resemble his job. He was the western regional director of the federal Office of Thrift Supervision.

Born out of the savings and loan crisis of the 1980s, the Office of Thrift Supervision was established in 1989 by act of the United States Congress. Headquartered on G street in Washington, DC, it occupied a multilevel concrete beige building, a very distant and hideous cousin of the strident, angular geometries of Brutalist architecture constructed in 1960s Britain[1]; its mandate to supervise, charter, and regulate the thrift industry. Thrift institutions consist primarily of savings banks and saving and loans associations. They historically formed to take deposits from consumers and use these funds to make residential mortgage loans—a traditional banking model, borrow short to lend long. By the end of 2007, the agency oversaw 831 thrift institutions with assets of $1.57 trillion, as well as 470 thrift holding companies with U.S. domiciled assets of about $8.5 trillion. It divided the United States into four regulatory regions, Northeast, Southeast, Central, and Western with associated provincial offices in Jersey City, Atlanta, Chicago, and Dallas. Each region had a regulatory director.

For over 30 years, Dochow rose steadily in the world of the government regulator. Ivy League educated, he had earned a master's degree in public administration from the John F. Kennedy School of Government at Harvard University, a master's degree in business from the University of Oregon, a bachelor's degree in finance from the University of Washington, and he was a graduate of the Pacific Coast Graduate School of Banking.

He began his regulatory career in 1972 with the Office of the Comptroller of the Currency and steadily worked up through the organization, becoming the assistant chief national bank examiner in Washington. In 1985, he moved to the Federal Home Loan Bank of Seattle as a senior vice president and director of supervision for thrift institutions and holding companies. He was quickly promoted to the position of executive director for the Office of Regulatory Activities for the Federal Home Loan Bank System in Washington, DC. In 1989, he joined the newly formed Office of Thrift Supervision. It was here, unknown and at the zenith of his career, that an extraordinary set of circumstances thrust him, unexpectedly, into the media spotlight.

On August 21, 2007, the Office of Thrift Supervision issued a press release. It went largely unnoticed by the financial broadsheets and investment media. The release named the new regional director of the west region as Darrel W. Dochow. After years of regulatory service, it must have seemed a fitting reward to him. Little could he have imagined it would end badly, with he and his superiors pursued by an angry crowd of investors, taxpayers, and politicians. If he had known this, that fateful day in August would have been considerably more somber.

But August 21, 2007, was a wonderful day for Dochow. With his rare mix of seasoned experience underpinned by a world-class education, few would have doubted, on paper at least, his choice. Indeed, so confident was the Office of Thrift Supervision in their man that the senior deputy director and chief operating officer, Scott M. Polakoff, excitedly proclaimed[2]:

> I am very pleased that our agency is able to appoint a highly capable leader of a key region at the OTS. ... Darrel's background and experience were important factors in his selection. We are fortunate to have such an outstanding individual as Darrel assume this position, particularly given his familiarity with many of the institutions in the OTS West Region.

The west region covered a huge area geographically; the states of California, Oregon, Texas, Mississippi, Washington, Arizona, Nevada, Colorado, Utah, Louisiana, Arkansas, Oklahoma, New Mexico, Kansas, Missouri, Idaho, Montana, Wyoming, Hawaii, Alaska, Northern Mariana Islands, and the territory of Guam. In monetary terms, it contained approximately half of all thrift assets across the entire United States of America (see Figure 2.2).

Dochow's government salary of $230,000 reflected the importance of the position. Endowed with comprehensive inspection and investigatory, surveillance, and compliance powers, he was a regulator with teeth, who could, if he so desired, bite, and hard. Part of the responsibility of any regulator is protection of the financial system from systematic risk, and this is why he or she needs sharp teeth. Dochow's role in this regard was to ensure thrifts in his region were operating in a market that was fair, efficient, and transparent.

Two tasks were particularly important for this to be effective. The first involved monitoring institutions to ensure that they were not taking excessive risks with depositor's money. The second was ensuring thrifts under his domain maintained a sufficiently large capital cushion to absorb losses should they arise.

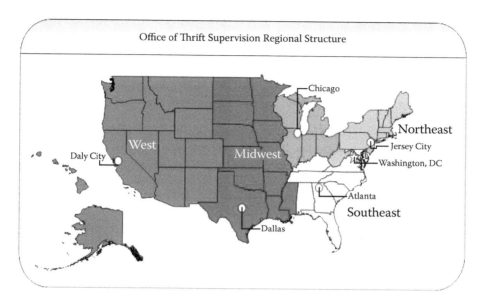

FIGURE 2.2
Map of the Office of Thrift Supervision (OTS) regions.

Headquartered in Pasadena, California, IndyMac Bank Corporation was one of the larger institutions under Dochow's watch. Its chairman and chief executive officer, Michael Perry, was a baby-faced, milky smooth peach-skinned fifth generation Californian, with straight brown hair, which was as thick and lush as he was intense. His eyes were large, nose small with a high forehead, and undersized rounded chin. Perry's oversized dark brown eyes flashed with arrogance and pride; pride at the independence of his company; independence to do with it as he pleased. It was not always so. Perry joined IndyMac in 1993 as chief operating officer when it had only four employees and it was owned by another financial institution, Countrywide Credit Industries, Inc. But in 1999, IndyMac gained independence, a fact Perry would boastfully point out to his staff at every available opportunity.

From IndyMac's inception as a savings association in July 2000, Perry grew the firm aggressively so that by 2006 it employed around 6,500 people and was the seventh largest savings association and ninth largest originator of mortgage loans in the entire United States. Perry's strategy for expansion was jejunely simple. Originate individual mortgage loans, bundle them together into securities, sell these securities as quickly as possible to investors on Wall Street, and repeat ad infinitum. The whole business model was predicated on the view that house prices would continue to rise perpetually. The key for IndyMac was focus on the so-called Alt-A loan market. These are mortgage loans issued to borrowers who have better credit than sub-prime borrowers but cannot fully document their income or assets. Alt-A loans are appealing to such borrowers because they charge a lower rate of interest than subprime mortgages.

Perry, in a breezy rush to construct his economic empire, bypassed the traditional route to growth—build a stable and loyal core deposit base of local customers. Instead, IndyMac fueled rapid expansion by reliance on costly Federal Home Loan Bank advances and the high-octane fuel of broker deposits.

Created by the United States Congress, the Federal Home Loan Bank consists of 12 regional cooperative banks that community banks, thrifts, commercial banks, credit unions, community development financial institutions, insurance companies, and state housing finance agencies can access to fund community lending. Around 80% of U.S. insured lending institutions are members of the Federal Home Loan Bank System. That IndyMac used Federal Home Loan Bank advances as a source of funding is not usual, for many financial institutions access these advances on occasion. But the extent of their use by IndyMac should surely have raised a red flag to the Office of Thrift Supervision. At the end of September 2006, IndyMac had over $9 billion in outstanding Federal Home Loan Bank advances. An examiner from the Federal Deposit Insurance Corporation pointed out these advances represented 34% of IndyMac's total assets, and suggested the situation should be monitored closely by the Office of Thrift Supervision. The Office of Thrift Supervision acknowledged these were "eye-opening stats."[3] But apparently it did little else. Almost 2 years later, in March of 2008, Federal Home Loan Bank advances remained persistently high at 32% of IndyMac's total assets.

IndyMac's loan production was explosive, up from $10 billion in 2000 to $90 billion in 2006. By the end of that same year, it had become the largest Alt-A lender in the nation with almost 18% of the market. IndyMac's stock price soared, and so did the acclaim. Between 2001 and 2002 it received the Inman News Services Most Innovative Use of Technology in Lending award and the prestigious 10X AWARD for its Internet delivery of mortgage products.[4] Perry received the Los Angeles Ernst & Young Entrepreneur of the Year award, the Advocate for Children Award,[5] and the Los Angeles Business Journal's Financial Industry Leader of the Year award. He was inducted into the Journal's Business Hall of Fame, and the American Homeowners Association chose IndyMac as its exclusive mortgage partner.[6] Perry and his senior management team, by offering loan products to fit a borrower's need, were lauded as enablers of the American Dream of home ownership. In less than a decade, Michael Perry had become a titan of the thrift industry.

It was then the trouble began. The exact date the discord started to sound is difficult to pinpoint precisely, perhaps as early as 2003. It began quietly at first, stories from insiders of loans being approved without any verification whatsoever of the borrower's income or assets, loans to borrowers with pitiful credit histories and questionable property appraisals. And then, beginning in 2006, property prices in IndyMac's largest markets—California and Florida—collapsed. It was not the gentle, gradual downward slide one might study in an economic textbook, nor was it a sudden 10% or so correction back to what real estate professionals call equilibrium or fundamental value. No, it was neither of those things. Instead, county-by-county, like a giant stack of cards, the housing market in California and Florida suffered a hair-raising downward descent. In Monterey County, California, from their peak, property prices fell a terrifying 75%.[7] Borrowers with no equity in their homes abandoned them by the legion; property developers declared bankruptcy and entire communities collapsed, starved of revenue generated by property taxes. The property bust sent shockwaves throughout the entire country, and eventually precipitated a financial panic, which in 2008 would engulf the entire world.

It was precisely at this moment when a crescendo of angry voices raised their most damning allegations; allegations that Perry's IndyMac had a deep-rooted apathetic culture of greed, which emanated from the very top, a culture that mercilessly sought to exploit the weakest and most venerable members of society, and a culture of unsound and abusive lending practices. Former *Wall Street Journal* reporter Mike Hudson documented a number of horrifying allegations, two of which stand out as unusually abhorrent[8]:

- An 86-year-old man, Mr. Ferguson, suffering from acute dementia solicited and made to think he was getting a better mortgage deal. He was not.* IndyMac directed more than $21,000 in fees to the broker, reward for *inducing Mr. Ferguson to take out a loan on terms much less favorable than were otherwise available to him.*

- A disabled United States Navy veteran promised he would be able to refinance his mortgage and payoff his credit cards and tax bills for a monthly payment of $526. The actual monthly payment was $631 or about 70% of his income. Further, the settlement charges were $5,962, almost twice the amount of $3,261 provided by the new loan.

Overall, Hudson concludes that IndyMac:

> … pushed through loans based on bogus appraisals and income data that exaggerated borrowers' finances; worked hand-in-hand with mortgage brokers who misled borrowers about their rates and other loan terms and stuck them with unwarranted fees; and treated many elderly and minority consumers unfairly.

An important duty of any financial regulator is to protect the public from being misled by fraudulent and abusive practices. In April and August 2001, July 2002, September 2003, November 2004, November 2005, and January 2007, the Office of Thrift Supervision conducted examinations of IndyMac.

---

* Also see INDYMAC: What Went Wrong? How "ALT-A" Leader Fueled Its Growth with Unsound and Abusive Mortgage Lending, 2008, Center for Responsible Lending.

At least once each year from 2001 to 2008 anywhere between 12 to 40 men in gray suits, Office of Thrift Supervision regulatory examiners, descended on IndyMac offices. Yet, despite the regularity of the examinations the Office of Thrift Supervision remained silent on this issue. But to those who were concerned and prepared to take on the responsibility that one imagines lay naturally with the Office of Thrift Supervision, it became clear very quickly that loans were being extended to borrowers who simply could not afford to make their payments.[9]

Up until the end of 2006, despite a rising tide of allegations, IndyMac delivered a strong performance. The average return on equity was 17% in 2004, 21% in 2005, and 19% in 2006. This was achievable, in part, because the secondary mortgage market had a voracious appetite for IndyMac's packaged mortgage loans. In practice, and unbeknownst to the investors, many IndyMac loans were extended with little, if any, review of borrower qualifications, including income, assets, and employment. Such practice is clearly fraught with difficulties associated with fraud and other abuse. Yet, it was baked into the *IndyMac way*. One of the operating procedures used by the thrift was the so-called *no doc*[10] in which income, employment, and assets were simply not required to be verified. There were also material and systemic weaknesses in the property appraisal processes used to support collateral on loans:

- A borrower requested a $3 million loan, reporting a self-employed income of $57,000 per month. The appraiser, in stark contradiction to IndyMac policy, was chosen by the borrower. The property appraised for $4.9 million. IndyMac did not inspect the property. The borrower made no payments on the loan before default. The property later sold for $2 million.
- A loan for $926,000 was approved for a borrower who stated a self-employment income of $50,000 per month. No attempt whatsoever was made to verify the borrower's assets. The appraised value of the property was $1.43 million. A total of $5,389 in payments was made by the borrower before defaulting on the loan. The property was subsequently listed for sale for $599,000.
- A property contained several appraisals ranging in value from $639,000 to $1.5 million. No documentation was put forward to support the higher appraisal over the lower. The loan for $1.5 million was approved. It defaulted.

In short, IndyMac engaged in unsound underwriting practices, yet the regulator remained silent.

Poor underwriting was not a serious problem for Perry and his team provided they could package the loans into securities and pass them onto investors in the secondary market. However, as the housing market recession deepened, demand in the secondary mortgage market began to dry up. For the first time in his career, Perry was faced with an economic environment that would test his leadership *mettle*. If he was to succeed, he would have to draw deep into the well of his intellectual reserves, store of good judgment, and courage and business acumen. Even then, Perry realized it would be a difficult battle. For if the secondary mortgage market shut down completely, drastic action would be required to keep IndyMac a going concern. If it was a temporary slowdown, maybe his firm could weather the storm. The question was, which was it?

As do all men who aspire to be great business leaders, Perry's eyes darted across the economic landscape, taking in all the financial and economic information he could. His brain, at full capacity, was calculating probabilities, profitability, and strategy. By January 2007, it was clear. Something big was taking place. A financial crisis the like of which few alive had experienced was looming. Both he and his senior management team had been lulled into a false sense of security during the boom years. They had never experienced a significant market downturn, yet here it was. Blind panic among his executive team would send shareholders scurrying for cover. Perry recognized this and tried to hold things together. Leadership was what was required, and leadership was what Perry desired to deliver. To demonstrate his credentials and establish a semblance of claim, he had to take decisive action. His mind flashed back to 1993, the year he joined IndyMac, when it was the submissive vassal of another institution. Then, inspired by his motivational mantra of *independence*, he acted.

In February 2007, Perry ordered the tightening of underwriting standards and slashed the maximum size of loans to home-builders by 75%. It was not enough. Two months later as the economic environment worsened, he ordered the firing of 400 employees; this he believed would save IndyMac $30 million per year. It was not enough. By August 2007, IndyMac was predicting a loss of $30 million for the third quarter alone. If true, this would be the first quarterly loss in the firm's history as a savings association. Economic conditions were deteriorating faster than he had anticipated. More decisive action was needed. But could Perry deliver?

Perhaps he should seek out a partner, a larger institution, which could infuse capital to support IndyMac through the crisis? But Perry's mantra of independence retarded him from pursuing this option to conclusion. His mind was so constituted that he thought more of what he might lose, than of what IndyMac might gain. He could not countenance the loss of his beloved IndyMac, he could not bring himself to relinquish his ability to do with her as he pleased. He needed something else, a new strategy. If not Alt-A, then what? And then it struck him, IndyMac must move away from its traditional core business of Alt-A mortgage origination to more conservative conforming loans.[11] The announcement was duly made on the 22nd of August 2007. At virtually the same time, with only 33 retail branch locations, Perry decided to make increased use of broker deposits.

Known as *hot money* in the industry, these deposits are extremely volatile. Brokers, who are looking for the highest yields, usually keep their deposit at the same bank for six months or less and then move onto another bank, which offers higher yields.

While Perry could not sacrifice IndyMac's independence, he would continue to sacrifice her employees, and in September, as financial conditions continued their downward spiral, plans to eliminate up to 1,000 jobs were announced. As November rolled in, the projected loss of $30 million for the third quarter turned into a catastrophic realized net loss of $203 million, almost seven times the August initial estimates.

At length there came a time which Perry must have dreaded, and which he had tried to keep distant—the time when he had to admit IndyMac could no longer pass on its neatly packaged bundles, of what the media now referred to as *toxic assets*, to investors. It was clear, Wall Street simply did not want to hold onto mortgage-backed securities, and the market was closed down completely. IndyMac had little choice but to keep on its books, loans it had intended to pass through to secondary market investors, loans it had extended to people who simply could not afford to pay, and loans on properties for which the appraisal process was questionable. In short, IndyMac had to eat its own cooking. It must have been at this time Perry first sensed his leadership metal had been tested and found wanting. He may already have glimpsed the disaster that lay ahead of him. If he did, it was not apparent to investors. But with no market for IndyMac's product, he resigned himself to placing the unsellable loans into the *held for sale portfolio*. It swelled to $10.7 billion by December.

As the festive month ticked slowly by, there were no bids and no market. The loans in the *held for sale* portfolio were transferred to the *held to maturity portfolio*. Perry's senior management team must have been terrified by this prospect, for they realized the music had stopped and they were left holding the *toxic waste*. Their only hope lay in the quick recovery of the secondary market. This may have been a weak and flimsy hope, but it was all they had. Perry and his team clung opportunistically onto it; IndyMac would survive with Perry at her helm. The optimism was misplaced.

For Dochow had begun to pull together his best regulatory team. Echoing Air Chief Marshal Sir Trafford Leigh-Mallory's "Big Wing Theory" repulsion of the German Luftwaffe during the Battle of Britain, Dochow corralled the largest team of men in gray suits ever gathered[12]; in all, 40 regulators from the Office of Thrift Supervision and three members from the Federal Deposit Insurance Corporation. Their orders were to complete a comprehensive inspection of IndyMac. As with the Royal Air Forces heroic defense of England, the element of both surprise and overwhelming force were essential. The men in gray suits, armed with their instructions, descended in huge numbers on IndyMac offices on January 7, 2008.

CAMELS[13] is a system used by the Office of Thrift Supervision to evaluate a thrift's overall financial condition. It results in an overall score, which ranges from 1 to 5. A score of 1 is the highest rating, representing strongest performance, risk management practices, and the least degree of supervisory concern. A score of 5 is the lowest rating, representing weakest performance, inadequate risk management practices, and the highest degree of supervisory concern. By the 17th of January 2008, the regulatory review was over. IndyMac lay fully exposed. One would have thought it was ripe for closure. But no, Dochow and his superiors hesitated. Instead of closure, they lowered the CAMELS composite rating from 2 to 3.

It was perhaps at this time when serious questions about the regulatory competence of the Office of Thrift Supervision began to surface. For within a month of the CAMELS downgrade, on February 12, 2008, another bombshell exploded. Perry, his oversized brown eyes now less intense, more fearful, announced a net loss of $509 million for the fourth quarter of 2007 and the suspension of common stock dividends. In an ironic twist of fate, the loss included a $600 million write down on the increasingly toxic assets in the held to maturity portfolio. Perry was eating his own cooking despite the foul taste. He might be able to swallow, but would his stomach be able to digest the toxic waste his company had been feeding investors? Blind panic by investors was perhaps the only rational response, but rationality requires informational transparency. IndyMac investors did not have this.

As April turned into May, the financial news grew grimmer. The held to maturity portfolio ballooned to $11.2 billion, and in a reflection of the[14] "unsafe and unsound manner in which the thrift was operated" defaults skyrocketed, over 12% of loans were 90 days or more in delinquency. IndyMac was slowly drowning in its own toxic waste. Yet, Dochow, with all his regulatory teeth did not act, maybe even could not act; and perhaps Perry grew to know this. For deep within the dismal discipline of economics there lays the little discussed theory of *regulator capture*. It explains why a regulatory agency created to act in the public interest instead acts in favor of the commercial interest it is supposed to regulate. There grows over time a very close relationship between the regulator and regulated. The regulated would like to capture its regulators so it can be free to do as it pleases; and the regulators themselves may desire capture, for *capture* can pay very well. Indeed, it is not unusual for regulators to leave their governmental job for lucrative positions in the sector they once oversaw. There is a revolving door between the regulatory agencies and Wall Street. Dochow knew this and so did Perry.

And then came some terrible news for Perry, news that could sink IndyMac immediately. The independent auditor, over which Perry could exercise little control, identified a number of adjustments that needed to be made to the March 31 financial statements. The effect would put IndyMac's capital ratio below the critical *well-capitalized* threshold of 10%. This was important because the use of brokered deposits as a source of funding was limited to well-capitalized institutions. IndyMac depended on broker deposits, which had risen from $1.5 billion in August 2007 to over $6.9 billion by the end of March 2008.

The Office of Thrift Supervision used four broad definitions in terms of capital ratios—well-capitalized, adequately capitalized, undercapitalized, and significantly undercapitalized. Falling below 10% would push IndyMac into the *adequately capitalized* category. But adequately capitalized institutions were required to obtain a waiver from the Federal Deposit Insurance Corporation in order to accept brokered deposits. Perry and his team did not want to be in the humiliating position of having to apply for such a waver—it would signal distress to the financial markets and the public. In addition, the independent auditor refused to sign off on their interim financial statements if management did not make the necessary adjustments. Terrified, Perry scrambled to gather his senior management team around him. What was he to do?

He knew, even with all his powers of charm, he could not influence the auditor, but perhaps, echoing the presidential election rallying cry of candidate Barack Obama's "audacity of hope," he could sway Dochow. "Get me Dochow," he must have screamed at his personal assistant, "and now!"

Sometime later, on the 9th of May, Dochow, Perry, and the auditor gathered their teams for an urgent telephone conference[15]:

> During the call, the CEO [Perry] asked if OTS [Dochow] would allow IndyMac to record a May 2008 capital contribution from IndyMac's holding company to IndyMac as of March 31, 2008. According to E&Y [the auditor—Ernst & Young] officials, the OTS official acknowledged the issue of the E&Y's proposed adjustments and agreed to IndyMac's proposal to backdate the capital contribution. As a result, IndyMac's total risk-based capital ratio was restored back over the 10% "well-capitalized" threshold for the March 31 report.

Dochow had crossed the line which separates regulator from regulated. The backdated capital contribution would allow IndyMac to be represented as a well-capitalized institution, when in fact, it was not well-capitalized at that date. Partnered in a close tango-like embrace, Perry and Dochow airily danced around the regulatory requirement that IndyMac obtain a waiver to accept brokered deposits. The signal of *distress* would not be transmitted to the financial markets or public. The invisible hand of capitalism would not be allowed to run its natural course. Emboldened by his success with Dochow, Perry announced to the world[16]:

> Given the decline in our stock price, some people have questioned IndyMac's survivability in the current environment. I am here to tell you that I believe we have turned a corner.

IndyMac's share price soared.

That the regulator should willingly engage in such deceit is outrageous, doubly so because of the bevy of complaints, weak financial standing, and poor business practices of IndyMac. But these complaints had all fallen on deaf ears. Dochow and the Office of Thrift Supervision simply refused to listen. The regulator had teeth, but would not, could not bite. Instead of protecting the flock from the wolves, it joined the pack in a savage assault. The theory of regulatory capture had once again been vindicated. With the quite scandalous assistance of Dochow, IndyMac retained a CAMELS rating of 3. The maleficent hand of the men in gray suits was hard at work.

Then on June 26th, an extraordinary event occurred; an event which would suddenly expose the disinterested Dochow, and at the same time give focus to the gaggle of agitated voices of complaint and concern. It began with a U.S. Senator from New York, Charles Ellis Schumer. With the U.S. economy weak and the stock market nervous after the collapse of the investment bank Bear Stearns, Schumer took the unprecedented step of writing a letter to the Office of Thrift Supervision. The letter distilled the longstanding voices of concern into a blow-by-blow account of issues the regulator should have been aware of and taken actions to correct. It suggested IndyMac was on the verge of failure.

The letter was leaked to the public. The response was immediate. In the three days prior to the release of Senator Schumer's letter, IndyMac had an inflow of deposits of $32.2 million. The day after release, Friday June 27, saw a net outflow of $4.5 million. This rose, as media outlets seized on the letter, to $78.2 million by Saturday, and $118 million on Sunday. In eleven business days, depositors, in a blind panic fearful of losing their money, withdrew more than $1.3 billion. The invisible hand had been allowed to work, IndyMac was finished, vanquished by a classic *run on the bank*.

Bizarrely, even in IndyMac's death throes, as it uttered its last gasps, Dochow and the Office of Thrift Supervision moved slowly. They continued to give it a high-composite CAMELS rating right up until shortly before it failed. Only on the 1st of July was it downgraded to 5. Ten days later, on the 11th of July, the men in gray suits once again descended upon IndyMac's corporate headquarters, this time to close it down. Its shares fell 57% on the news. They turned out to be worthless.

Like the captain of some stricken luxury liner, Perry remained *on deck* until the very last moment. He was at IndyMac's corporate headquarters to greet the regulators, a massive team of 130 men and women from the Federal Deposit Insurance Bank failure division. The company he had grown from a handful of people to over 6,000 had crashed down around him; busted, destroyed, Perry had led it to utter and total ruin. With IndyMac's demise, a titan of the industry, Michael W. Perry, CEO, had fallen.

The collapse of IndyMac was one of the largest bank failures in American history. It cost the Deposit Insurance Fund $10.7 billion dollars. As is often the case in the immediate aftermath of such a calamity, the finger of blame was pointed in totally the wrong direction—at Senator Schumer. Indeed, in an unusually terse statement, which must have been signed off by Dochow and the agency head, John Reich, the Office of Thrift Supervision stated boldly[17]:

The immediate cause of the closing was a deposit run that began and
continued after the public release of a June 26 letter to the OTS and the
FDIC from Senator Charles Schumer of New York. The letter expressed
concerns about IndyMac's viability.

But as the dust settled, it became clear to all who cared to look, that the
maleficent hand of the men in gray suits had been at work. For it was quickly
discovered that senior management across the board at the Office of Thrift
Supervision had authorized backdated capital at other thrifts on other occasions. It appeared to be common. In a blistering critique of this perfidious
practice, the Office of Inspector General scolded[15]:

> We consider these matters very serious and find it alarming that such
> high level OTS officials were not only aware of the backdating at two
> thrifts, but either directed or authorized the thrifts to backdate the capi-
> tal contribution. Approving or directing the thrifts to backdate these
> contributions is inappropriate as the accounting treatment is not in
> accordance with generally accepted accounting principles (GAAP) and
> allows for misleading financial reporting by the thrifts.

Then came another shocking discovery; Dochow, two decades earlier, had
been demoted for being a *zombie* regulator—allowing an insolvent institution to remain open. He was at the time the head of supervision and regulation at the Federal Home Loan Bank Board, and he delayed and impeded
proper regulation of notorious businessman Charles Keating's California-
based Lincoln Savings and Loan.[18] It collapsed in 1989 with losses of $3.4 bil-
lion and 23,000 investors lost close to $300 million. Keating was convicted on
fraud, racketeering, and conspiracy charges, and sentenced to the maximum
of 10 years in prison. Dochow was demoted for his duplicity and sent to a
regional office. One naturally wonders why he was not fired.

With the failure of IndyMac, Dochow's artifice caught the attention of the
mass media. It was portrayed as a grotesque enactment of the Pied Piper of
Hamelin in which the familiar roles of humans and rodents were reversed.
A rat, symbolic of the regulator, played the tune, while the people, a crowd
of innocent investors and homeowners, followed eagerly behind unaware
they were being led to their doom. The public, politicians, and media commentators became enraged. "Surely," came the angry cry, "The role of the
regulator was to supervise banks not conspire with them?" The question
was asked again, and again, each time, as new facts became known, with
growing venom.

Vilified by the press, there could be no reprieve for Dochow. He was unceremoniously relieved of his duties as western regional director in December
of 2008 and quietly retired in February 2009.

Investors, taxpayers, and politicians were outraged by the actions of
the Office of Thrift Supervision. For there was little room for any doubt

whatsoever, the regulator was riddled with the noxious twin maggots of incompetence and unconcern. What once had been designed to serve the public interest now had rotted to a malignant core of dilatory enforcement, moronic judgment, and senior managers with a conspiratorial mindset and dubious track records. Unreason had taken hold and choked out the *good sense* from an entire regulatory agency. In an attempt to cut out its rabid growth, John Reich, the head of the agency, was asked to step down. He did so in February 2009. Scott Polakoff's gushing words of praise for Dochow, "We are fortunate to have such an outstanding individual as Darrel assume this position..." asphyxiated his regulatory career. He was placed on administrative leave in March 2009 and retired from his position of senior deputy director and chief operating officer in July of that same year. But by then, the damage was done. The entire agency was closed in utter disgrace during October 2011, a shameful and sorry end for an agency founded in a once noble cause.[19]

The Apathetic Regulator

For a moment, there was tense silence in the courtroom, which was suddenly broken by Miriam Siegman's rising voice[20]:

> The man sitting in this courtroom robbed me. In an instant his words and deeds beat me to near senselessness. He discarded me like road kill. Victims became the byproduct of his greed. We are what is left over, the remnants of stunning indifference and that of politicians and bureaucrats. Six months have passed. I manage on food stamps. At the end of the month I sometimes scavenge in dumpsters. I cannot afford eyeglasses. I long to go to a concert, but I never do. Sometimes my heart beats erratically for lack of medication when I cannot pay for it. I shine my shoes each night, afraid they will wear out. My laundry is done by hand in the kitchen sink. I have collected empty cans and dragged them to redemption centers.

Bernard Madoff sat bolt upright in his seat. The victim's words had pierced through his usual unemotional mannequin form. For an instant, the enormity of his crime rose up in terrifying form before him. Then, he lurched back into his seat, his signature smirk creeping slowly across his aged face. It was as if he had momentarily forgotten and then remembered his guilty plea; Miriam Siegman's words could do him no more harm than he had already done to himself. But her hoarse cry of "stunning indifference" aimed at regulatory bureaucrats, resonated deeply with that coterie of Madoff victims ruined beyond redemption. This time the regulator under fire was the Securities & Exchange Commission (SEC).

The SEC was born in 1934 in the aftermath of the Wall Street Crash of 1929, the collapse of Ivar Kreuger's match making empire, numerous other corporate scandals, and abusive trading practices. Its core mission is to protect investors; maintain fair, orderly, and efficient markets; and facilitate capital formation. It is the preeminent protector of the financial markets in the United States of America. Yet, time and again in that packed courtroom, the anonymous bureaucrats of the SEC, those unknown men and women in grey suits, found themselves the subject of blistering criticism.

> In addition to Madoffs actions, our own government has failed us completely. The failure of the SEC to act when they had all the information necessary to stop Madoff in his tracks.

> Hearing from the SEC that he was a safe broker, we thought we were OK with leaving our money with him. We now have nothing.

> We trusted the SEC to protect us and they failed us.

> The SEC and other regulatory agencies were constantly assuring investors that Bernie ran an honest business. The SEC did my "due diligence."

> All the years we had our money he was robbing us and we had no idea. The SEC which was supposed to protect us did not do their job.

What these hapless victims could not have known, but had experienced individually and collectively in their ruinous financial losses, was the staggering asinine stone-deaf impassivity by which SEC regulators conducted their agency's business. Take for example the simmering controversy surrounding a 37-year-old mother of two by the name of Meaghan Cheung. She was a medium, spare woman, slightly stooped, with a shallow, pale forehead, prominent angular nose, and dark adroit eyes accompanied by shoulder length, deep brown, almost black hair. Her mouth was perhaps her best feature, for, while the lips were pinched, they had a kind of cold refinement, which complemented the angry intensity of her stare.

It was early January in New York City—2009. A cold, still winter day, where expelled breath dances and swirls in a light gray cloud, rises upwards and disappears forever into the city ether. Madoff, weeks earlier, had been exposed as a total fraud. The question of how he had been able to get away with such a massive scam for such an extended period of time was beginning to raise its head. Cheung emerged out of the ashes of the Madoff collapse as a real life caricature of Inspector Clouseau, the bumbling incompetent police inspector of the French Sûreté.

Cheung was standing on the steps of her luxury Manhattan apartment, dark angry eyes staring menacingly out from behind thick, framed acetate glasses. On her head, a sky blue tam matching almost exactly the shade and tone of the English Football Association's 1987 Cup winners—Coventry City. Her left hand wrapped tight in a matching glove clutched forcefully onto a used paper coffee cup. She wore a blackened padded jacket, which combined with the slight stoop and snarling facial expression gave her the look of a demented bag lady. But Cheung was no vagrant. Unlike Miriam Siegman, who was reduced to desperate acts of beggary by the gluttonous thievery of Bernie Maddoff, Cheung could spend as she pleased. Her luxury Manhattan co-op apartment was proof positive of this. She was the former SEC branch chief of the New York enforcement division, and notorious as the SEC regulator who shut down a fraud investigation into Madoff 3 years before his voluntary confession.

Her snarling expression was directed at a reporter. Fingers were being pointed at her. Questions asked about her role in the Madoff scandal. Why had she closed the investigation? Why had she failed to uncover the scam? Questions one would have hoped she would have been eager to answer. But Cheung had studied hard to graduate from Yale and then Fordham University Law school. And she was well-schooled in civil servant "no speak." She would not answer, would not explain. Instead, she went on the attack yelling defiantly,[21] "There's nothing I can say about what we did in this investigation other than to say we worked as hard as we could." But for a man by the name Harry Markopolos, the reporter might have been satisfied with her response.

Markopolos contacted the SEC in 2000, 2001, 2005, 2007, and 2008 explaining each time in excruciating detail how and why Madoff's business was a giant Ponzi scheme. In November 2005, Markopolos laid out the case against Madoff to Cheung. In January 2006, Cheung authorized an enforcement investigation into Madoff. However, mirroring Inspector Clouseau's inability to identify crime even when it is being committed in front of him, Cheung and her team found "no evidence of fraud."[20] She recommended the case be closed. And this, even though Markopolos had explained in great detail how the fraud was being conducted, where to look, and what questions to ask. Markopolos had given Cheung a roadmap and flashlight, but like blundering Clouseau, she failed to follow it to the correct destination.

Harry Markopolos, in his testimony before the U.S. House of Representatives, expressed his frustration at her inactions thus[22]:

> She [Cheung] never grasped any of the concepts of my report, nor was she ambitious enough or courteous enough to ask questions of me. Her arrogance was highly unprofessional, given my understanding of her responsibilities and mandates.

Perhaps, one might plausibly reason, Cheung was useless rather than venal, her ineptitude a rare extreme, an unusual aberration, she little more than a solitary rotten egg. But this still begs the nagging question how had she climbed to such a prominent position within the SEC? How could somebody who "never grasped any of the concepts" of a basic Ponzi scheme be in charge of an investigation into one? And then there were the questions raised by the victims, questions that hinted to more than the deficiency of an individual, but to a wider systematic failure within the SEC.[20]

> According to the SEC Inspector General, 27 SEC employees were involved in 7 investigations of Madoff over 11 years, all of which found no evidence of fraud.

> Prior to investing my husband and I researched Madoff and found the 1992 SEC report that assured us his was a legitimate business and that he was a respected professional with a top reputation.

> I was aware that Madoff was investigated by the SEC on several occasions, and subsequently, no indiscretions were found. I felt secure by the SEC's findings and that my investments were SAFE.

> We were devastated by the SEC's failure to uncover Madoff's fraud and its continued stamp of approval bestowed on Madoff over the decades of his crime.

And then a quite astounding fact was uncovered. Shana Madoff, Bernie Madoff's niece and the compliance director at her uncle's defunct firm, had married SEC regulator Eric Swanson. Attending the wedding was regulatory bigwig Lori Richards, the director of compliance inspections and examinations at the SEC. California Congresswoman Jackie Speier summed up the shock at this revelation: "Mr. Swanson was the lead attorney on this case. He leaves the SEC, marries Mr. Madoff's niece."[23]

As each noxious revelation seeped out, it became imperative for the SEC to come forward and explain. That opportunity eventually came on February 4, 2009 at a hearing before the subcommittee on capital markets of the U.S. House of Representatives.

A sense of quiet expectation hovered over the packed committee room. At last, Madoff's victims and the public were going to get answers. For Madoff himself, up until that time, had remained tight lipped. The gathered politicians, spurred on by the media frenzy, dispensed their questions with the precision and accuracy of the British Army's famed L115A3 sniper rifle. It began with the basic, but critical question from New York Congressman Gary Ackerman.

> You took action after the guy confessed. He turned himself in. ... Why didn't you find him, is the question. ... How did you screw up?

This was followed, rapid fire, by the member of Congress from New York, Carolyn Maloney:

> Mr. Markopolos in his testimony earlier testified that he brought complaints 5 times in writing to the SEC, and these were detailed complaints.

It wasn't, "I think something's wrong." These were detailed complaints that this is wrong. They are not trading ... how many more times would a whistleblower have to bring complaints to the SEC for them to have investigated the Madoff case? ... One of the things he said was that Madoff wasn't conducting trades. Now, if you went in and just asked for the trade slips ... then you could have shut him down in one half hour. You could have shut Madoff down in one half hour by just following up on one of his allegations that they were not conducting trades. ... He [Mr. Markopolos] offered to risk his life to work with the SEC to prove this fraud. Why was that request turned down?

Yet another obvious question was raised by Democratic Congressman Joe Donnelly:

... one of the [red] flags was a $50 billion fund with a one person account-ing firm. Why was that one flag not enough for you to shut down earlier?

Upstate New Yorker and Blue Dog Democrat Mike Arcuri followed up with further quizzing:

... if you saw an investor who was giving 4 percent of the profit he would normally receive to the feeder companies, would you think that might be a red flag? ... How about if you saw a company that continually at the end of each period turned their cash into government securities; would you consider that perhaps a red flag? ... you had the scenarios that I just described; you had a credible lead, and yet nothing was done by the SEC, correct?

The penetrating questions came from both sides of the house, Democrat and Republican, standing shoulder-to-shoulder demanding answers. Republican Congressman Bill Posey summed up the consternation at the inaction of the SEC:

It is just such an incredulous tale, I think, for everybody up here to understand ... it is just incredibly unbelievable to the people in this committee to hear these stories. ... Besides Mr. Markopolos, we have the Barrons's article, we had what is called the Ocrant article. We had Merril Lynch, Goldman Sachs telling their investors, don't touch this. This is impossible. This has got to be a scam. We have hedge fund managers, money managers with the absolute minimal amount of due diligence, you know, telling their clients by the thousands, this is a joke. This can-not possibly be working. Stay away from this thing. And yet, you know, our enforcement agency is blind to the whole—I mean, it is literally hard for everybody to believe.

Subjected to this piercing examination were a bevy of the SEC's senior leadership including Linda Thomsen, SEC director of SEC enforcement, Erik

Sirri, SEC director of trading and markets, Lori Richards, director of compliance inspections and examinations, and Acting General Counsel Andrew Vollmuer. The audience in the packed committee room expected clear and concise answers to the politicians' questions. Instead, media, public, and Madoff victims were subjected to a nauseating dose of civil servant no speak. The exchange between California Congresswoman Jackie Speier and Linda Thomsen of the SEC captures the spirit of the SEC's unresponsiveness.

Speier: "All right, I want you to each grade the SEC on how they handled the Madoff case, very quickly."
Thomsen: "Can't do it."
Speier: "Why can't you do it?"
Thomsen: "Because it would inevitably – "
Speier: "You are giving an opinion. Did the SEC do a good job? A, B, C, D, or F?"
Thomsen: "I wish we had found it earlier."
Speier: "Would you give the SEC an F?"
Thomsen: "I would not. I would not grade it."

And on and on it went. As the tactic of the SEC officials became clear, the gathered congressmen were riled and exploded with a voluminous, choleric verbal assault. New York Congresswoman Carolyn Maloney exclaimed:

> I find this absolutely outrageous, and if you won't answer it, I think I am going to appeal to the chairman to subpoena and find out what you did in this case.

Paul Kanjorski, chairman of the subcommittee, scolded:

> How do you explain the fact that you not only missed this, but now that Congress is attempting to close the loopholes and attend to it, you feel disposed not to cooperate 100%. ... You cannot help if it is a pending criminal investigation; you cannot help if the Inspector General is doing something; you cannot help if it is an ongoing violation. I mean, if there is a snowstorm in Washington, the SEC cannot help.

Even the harsh words of Congressman Gary Ackerman failed to spur the men in gray suits to be more forthcoming:

> I am frustrated beyond belief. We are talking to ourselves, and you are pretending to be here. ... You have told us nothing, and I believe that is your intention. I figured you would leave your blindfolds and your duct tape and your earplugs behind, but you seem to be wearing them today. What the heck went on? ... one guy with a few friends and helpers discovered this thing nearly a decade ago, led you to this pile of dung that is Bernie Madoff, and stuck your nose in it, and you couldn't figure it out. You couldn't find your backsides with two hands if the lights were

on. Could you explain yourselves? ... You have totally and thoroughly failed in your mission. Don't you get it? ... How did you screw up? ... We thought the enemy was Mr. Madoff. I think it is you. You were the shield. You were the protector. And you come here and fumble through make believe answers ...

Finally, once time had almost run out, Republican Congressman Bill Posey, exasperated by the day's proceedings exclaimed:

I haven't seen this much bobbing and weaving since Muhammad Ali's rope a dope.

Ali, perhaps the greatest boxer of all time, had used the "rope a dope" in his legendary *rumble in the jungle* fight against bone crushing heavyweight George Foreman. Ali's strategy involved "resting" on the ropes of the boxing ring, taking blow after blow to nonvital parts of his body, while defending his head. By round eight, heavy-hitting Foreman was spent. Ali, light-footed and fresh, delivered a series of lightning fast punches, knocking Foreman clean out. And thus in the *battle* with Congress on that February day in the winter of 2009, the SEC's rope a dope strategy was executed with the precision, skill, and timing not seen since the dominance of heavyweight boxing, some 30 years earlier, by the great Muhammad Ali. The SEC, through its unresponsiveness and inaction, had delivered yet another *black eye* to the unfortunate victims of Madoff.

In truth, the maleficent hand of the men in gray suits is not much discussed among investors; and this is a real shame. For it is only in dire circumstances that investors think about and reach out to the regulator for support. Alas, and all too often they find it is not a beneficial hand onto which they have grasped but its maleficent cousin dressed in the identical suit of gray. Rather than an open extended palm guiding them to safety, they find themselves subject to a clenched iron fist intent on delivering to them yet another devastating financial blow; and this may come as a great surprise. Yet, the truly amazing thing about the unseemly incidents discussed in this chapter is that anyone can recognize the maleficent hand of the men in gray suits if they care to look. It is so blatant, so obvious, that there could be no confusing it. The real mystery is how cardboard cutout regulators are able to survive and thrive. That they do, and in large numbers, is perhaps, further validation of Konrad Lorenz and Niko Tinbergen's 1937 experiment. The inane squawking of those long, forgotten turkeys echoes across history to warn us that unreason abounds in places where it must not.

Endnotes

1. Renowned British architects who dabbled in this style include the likes of Peter and Alison Smithson and the utterly disgraced John Poulson. Poulson's greatest work in this style is perhaps the Leeds International Baths building, constructed in 1967 and located in the county of Yorkshire, England.
2. For further details, see Office of Thrift Supervision, Press Release: "OTS Appoints Darrel W. Dochow West Regional Director," August 21, 2007. (www.ots.trea.gov/_files/777059.html).
3. See Office of the Inspector General (2009a), Safety and Soundness: Material Loss Review of IndyMac Bank, FSB, Audit Report, OIG-09-32, Washington, DC.
4. The 10X AWARD is given to a company, product, or technology application having an exponential impact on mortgage lending. See Mortgage Technology (http://www.mortgage-technology.com/) for further details.
5. This is a prestigious award given to those who demonstrate an exceptional commitment to children and their welfare. It is presented by Ettie Lee Youth & Family Services, which is named after Ettie Lee, founder, who is renowned for her work with children.
6. The American Homeowners Association's mission is to serve homeowners and first-time home buyers by expanding their purchasing power, helping them make more informed decisions, and representing their key interests in both the public and private sectors. See American Homeowners Association press release (http://www.ahamembership.com/press/press_080802.cfm) and American Homeowners Association (http://www.ahamembership.com) for further details.
7. By the end of May 2009, residential house prices in Monterey County, Monterey Region, High Desert, Santa Barbara County, Riverside, Palm Springs, Central Valley, and Sacramento had declined from their peak by over 50%.
8. Center for Responsible Lending (2006).
9. See, for example, Office of the Inspector General (2009a), Safety and Soundness: Material Loss Review of IndyMac Bank, FSB, Audit Report, OIG-09-32, Washington, DC; and Center for Responsible Lending (2006), IndyMac: What Went Wrong? CRL Report (www.responsiblelending.org).
10. This is short for "no documentation loan," also known as "liar loans." The applicant is expected to supply their name, address, Social Security number, and contact information of the employer if any. The underwriter determines the amount of the loan using the applicant's credit history, appraised value of the property, and size of the down payment. Also see Office of the Inspector General (2009a), Safety and Soundness: Material Loss Review of IndyMac Bank, FSB, Audit Report, OIG-09-32, Washington, DC.
11. And reverse mortgages. Conforming loans, purchased by Freddie Mac and Fannie Mae have rigorous underwriting and other standards.
12. For a review of IndyMac.
13. CAMELS is an acronym used by United States Financial Regulators. It is derived from the six areas of a regulatory exam, Capital adequacy; Asset quality, Management and administration; Earnings; Liquidity; and Sensitivity to market risk. The CAMELS acronym comes from the first letter of each of the six dimensions.

14. See Office of the Inspector General (2009a), Safety and Soundness: Material Loss Review of IndyMac Bank, FSB, Audit Report, OIG-09-32, Washington, DC.

15. See Office of the Inspector General (2009b), Safety and Soundness: OTS Involvement with Backdated Capital Contributions by Thrifts, Audit Report, OIG-09-037, Washington, DC.

16. See Tom Petruno, Shares Rise on a Predicted Return to Profitability for the Lender, Though Doubts Remain, *Los Angeles Times*, May 2, 2008.

17. Press release issued by the Office of Thrift Supervision, 6:00 p.m. EDT on Friday, July 11, 2008, reference OTS 08–029. See Office of Comptroller of the Currency (http://www.ots.treas.gov). This was the very day IndyMac was closed.

18. See William Heisel, U.S. Banking Official Darrel Dochow Retiring after Furor over IndyMac Failure, *Los Angeles Times*, February 21, 2009.

19. On July 21, 2011, the OTS became part of the Office of the Comptroller of the Currency (http://www.ots.treas.gov/?p=consumercomplaintsinquiries).

20. The case is *U.S. v. Madoff*, 09-cr-00213, U.S. District Court for the Southern District of New York (Manhattan).

21. See Lorena Mongelli and Dan Mangan, The SEC Watchdog Who Missed Madoff, *New York Post*, January 7, 2009.

22. Testimony of Harry Markopolus before the U.S. House of Representatives Committee on Financial Services (Wednesday, February 4, 2009, 9:30 A.M.).

23. See *Hearing Before the Subcommittee on Capital Markets, Insurance, and Government Sponsored Enterprises of the Committee on Financial Services*, U.S. House of Representatives, 111th Cong., First Session, (February 4, 2009).

3

The Unpalatable Truth about Risk Management

That unreason is the even eviler twin brother of greed and that it exists in places where it should not provides a natural and urgent motivation for active management of risk. However, the *professional* discipline of investment risk management has struggled to live up to the task. Investment risk management has long been steeped in myth, magic, secrets, and legends. Anecdotal stories about the downfall of dastardly risk managers who *thwart* the actions of well-intentioned traders seeking nothing more than to make an *honest* profit are legion. The frequently recounted tale of *"Dr. Drewzinskoff and Mr. Mumbo"* captures much of the mythical lore.

We all gathered around, eager to hear the news—Dr. Drewzinskoff, the global head of risk management, had been replaced. *"We have a new head of global risk,"* announced the chief investment officer (CIO). A huge cheer went up from the gathered crowd—for Dr. Drewzinskoff was not well liked. There was something about him, which many found rather distasteful; something perhaps in the combination of his caustic intelligence, mousey Polish accent, and years steeped in the bowels of risk management which gave to him, somehow, a rather mournful aura. *"We now have as our head of risk an MBA[1] graduate!"* continued the CIO, signaling for calm, *"and a personal friend of the President. I would like to introduce him to you. Gentlemen..."* He paused, as an anticipatory silence fell over the gathered crowd, *"Here is Mr. Mumbo ..."*

And there he was, a squat little man, no more than five feet tall with protuberant fish-like eyes and a greedy smile creeping slowly across his apple shaped face. As the gathered crowd gazed at him in respectful silence he hopped, leprechaun-like, to his tiny feet. Slowly, ever so slowly, he began to walk to and fro, to and fro. One minute passed, two minutes, three, four... The entire room was hushed, transfixed by this pacing pixie. Suddenly and without any prior warning, Mr. Mumbo turned away from the gathered crowd and in an unnecessarily dramatic flourish began dancing rapidly backwards, somewhat like a Lilliputian Michael Jackson. At the very last instant, he swung around to face the startled crowd. Raising his elfin arms out wide, gesturing for all to take note: what he was about to say was going to be important, very important. *"I have a plan for this department,"* he boomed in a voice which belied his size. *"For risk measurement, risk modeling, and risk management. It is a large plan, a huge plan, an ..."* He paused, apparently searching for a more grandiose business school word. *"AN ENORMOUS INTEGRATED ENTERPRISE WIDE RISK PLAN!"* he bellowed triumphantly.

Nobody spoke, not even "loud mouth" Harry the repo trader. The CIO, eyes wide open, face flushed an unhealthy mixture of crimson and blue, stood quite still staring in bewilderment at Mr. Mumbo. But all he could manage to splutter was *"Friend of the President."* There it is, I thought to myself, Mumbo's big plan, a mumbo jumbo plan! I chuckled out loud as my heart sank at the thought of working for this dancing, prancing clown. Fortunately, Mr. Mumbo with his acrobatic theatrics and business school babble did not last long. Many years later, through the grapevine, I learned he had returned back to university to study law—somebody had informed him there was more money to be made in investment litigation than risk management, and you know what? They were right!

A Rather Vulgar, But Common, Perception of Risk Management

The unlamented departure of a despised risk manager, such as Dr. Drewzinskoff, and his replacement by an MBA with the intelligence of a baboon captures politely the rather vulgar perception of risk management held in many quarters. For on the trading floor and in executive offices, the risk management function is frequently characterized as a bureaucratic destroyer of value, an unnecessary consumer of scare corporate resources, and a total and utter waste of time. At the infamous energy company Enron, where rapacious greed fueled looting, bribery, and deception on an unprecedented scale, risk management served useful only as a conduit behind which to hide malodorous activity. The attitude of Enron's senior management toward it can be described at best as contemptuous[2]:

> Says a former Enron managing director, "... I treated them [risk managers] like dogs, and they couldn't do anything about me. ... I told my guys to f**k 'em."

The narration by Lewis[3] of his encounter with a chief risk officer reflects, in a more elegant fashion, the lingering perception held by a significant number of highly educated and skilled investment professionals:

> Sitting on the very tip of my chair, feigning interest in the mumbled string of motivational buzz words spouting out of the mouth of an unusually dull director of global risk, it occurred to me that if I looked hard enough, through the gray mist of the incoherent muttering, there would emerge some shape, some form to their ideas [about risk management], which as yet my colleagues and I could not perceive. I mused on

this thought, toyed with the idea of developing a statistical algorithm that would filter out the noise, revealing the underlying structure. My jocose thoughts were shattered by what was supposed to be the motivational crescendo—we all rose to our feet and clapped our hands somewhat like well-fed seals at feeding time at the local zoo, that is, with not much enthusiasm. Unfortunately, for that individual, there was no form to his ideas, no shape to his plan.

If risk management is so useful, goes the cry, why did it not predict the financial calamity of 2008? How was it that Bear Stearns, Lehman Brothers, and IndyMac, with all their teams of risk managers, many certified by the growing body of *professional* risk management associations, failed to spot and avoid the largest financial implosion since the Great Depression? Scores of risk managers found themselves alongside everybody else caught up in a financial whirlwind in which they witnessed the total collapse of the leverage loan market, evaporation of liquidity in the asset backed commercial paper market, and the failure of the subprime mortgage market. Between 2008 and 2009, in the United States alone, almost 8.5 million jobs were lost, capacity utilization contracted, and the money multiplier fell to historical lows.

The United States had spiraled into what became known as the "Great Recession." The utter and total failure to identify and mitigate the risks that precipitated the crisis exposed the impotence of many risk management functions and served to further undermine confidence in the value of this nascent profession. It remains the case that risk management, for many, is seen in no better a light than the discredited American economist Irving Fisher, who declared immediately prior to the Wall Street Crash of 1929 that "stock prices have reached what looks like a permanently high plateau." The crash, when it arrived, was pronounced by Fisher as "only shaking out of the lunatic fringe." Recovery, he asserted, was just around the corner—it was not. That Fisher was a vocal supporter of eugenics, which at the time espoused forced sterilization of "feebleminded" Americans; and that he clung rather foolishly onto Henry Cotton's incongruous theory of mental illness[4] resulting in the brutal butchering and lingering death of his beloved daughter, did little to help his cause.

Unfortunately, the repute of risk management has not been helped by the prevailing impression that risk management functions are staffed by bespectacled nerds, easily frightened lackeys who shake in uncontrollable fear at the sight of their own shadow[5]:

> Risk management tends to attract people who are not alpha males and alpha females. Trading tends to attract more aggressive personalities [who] can intimidate people who are more analytically minded. ... Hence, if a risk management pro tries to confront a trader about the workings of a deal, the latter hardly hesitates telling a risk manager to "get outta here," then trade the deal the way he or she likes ...

Unfortunately, this caricature bears more than a passing resemblance to reality. Take for example the debacle at Enron where the risk assessment and control (RAC) department was headed up by a bespectacled individual by the name of Rick Buy[2]:

> Rick Buy was a pleasant, paunchy man with glasses—a soft spoken sort, uncomfortable with confrontation. When his [risk] analysts raised issues with a deal, Buy would dutifully take them up the chain of command. But in a head-to-head with the company's senior traders and originators, it was no contest, as those on both sides of the table recognized. ... There were times when frustrated RAC executives refused to sign off on a bad deal, but Buy would overrule them. In 1998, John Hopley, who served for four years as one of Buy's top deputies, opposed a deal [to invest $20 million in a bankrupt British company]. ... He [John Hopley] refused to sign the DASH [deal approval sheet]. So Buy signed it instead. Three months after signing the agreement, the company went into liquidation, and Enron wrote off its entire $20 million investment.

Integrity, intelligence, industry, and testicular fortitude are essential qualities of a risk manager. One would hope those charged with heading up the risk management function are neither fearful nor willing to grant favors to anyone. Yet, in practice, the repute of the risk manager remains the subject of widespread ridicule and myth. The satirical story of the "Balloonist and the Man on the Ground" is perhaps one of the more family-friendly characterizations[6]:

> A Balloonist was flying in a hot air balloon and realized he was lost. He reduced altitude and spotted a Man On The Ground below. Lowering the balloon yet further, he shouted: "Excuse me, can you tell me where I am?"
> The Man On The Ground explained, "You are in a hot air balloon, hovering 10 meters above this field."
> "I can discern that you are a risk manager," bid the Balloonist.
> "Indeed I am ..." replied the Man On The Ground. "How did you know?"
> "Well," offered the Balloonist, "everything you have told me is technically correct, but it's of absolutely no use to anyone."

This allegory highlights the difficulty the profession continues to have in establishing itself as a value proposition within the corporate hierarchy. Risk managers, who are not risk takers, are perceived as trying to tell *moneymakers* what risks they can and cannot assume. The risk management function is seen as a cost, not a source of revenue. It is the risk takers who generate revenue, yet it is the risk management function that seeks to block or alter lucrative deals. This fundamental friction leads Tippins to lament[7]:

> In the real world, when a risk manager is the only person from the profession employed by the firm (or one of a very small group, as is the case with most large firms) it is hard to find respectability and camaraderie.

When few people in an organization understand the job title and duties of a risk manager it is easy to be underappreciated. Most risk managers do their job in a state of relative professional isolation.

The Emperor of Risk, His Lyre and the Palatine

Newspaper headlines, evening news segments, and even Hollywood block-buster motion pictures[8] highlighting the sycophantic character of prominent risk managers, have done little to quell the widespread perception of it as an inherently nugatory profession. In September 2006, Amaranth Advisors LLC, a multistrategy hedge fund based in Greenwich, Connecticut, lost around U.S. $6 billion, or two-thirds of its assets. The losses occurred in just under a month. Three months later, Amaranth funds were being liquidated. It was at the time the largest hedge fund collapse in history. Since Amaranth was supposed to be a diversified multistrategy hedge fund with strong risk management infrastructure, the sheer size of the losses took investors and media commentators by surprise.

> Amaranth billed itself as a "multistrategy" hedge fund. But its exposure ... suggests it was over-reliant on a single strategy to earn the more than 20 per cent annual returns it was making for investors before its collapse. (*The First Post*, September 20, 2006).[9]

The apparent reliance of Amaranth on a single strategy brought into sharp focus the issue of risk management. On March 29, 2007, the San Diego County Employees Retirement Association (SDCERA), which suffered losses in excess of U.S. $150 million, filed the first lawsuit against Amaranth for securities fraud, gross negligence, breach of contract, and breach of fiduciary duty, claiming that[12]:

DID INVESTORS IN AMARANTH ADVISORS LLC KNOW?

Amaranth, a coarse herb-like group of plants, is a very interesting name indeed for a hedge fund. In ancient Greece, through Roman times right up until English poet John Milton's epic poem "Paradise Lost"[10] it had been associated with longevity and immortality. The six or so years Amaranth Advisors existed serves as a poignant reminder not of immortality but of the dismal fate that awaits many hedge funds.[11] Today, *Amaranth* is more commonly known as *Pigweed*, an invasive pest fit only as food for swine.

In truth, the fund, against its own espoused investment policies, effectively operated as a single-strategy natural-gas fund that took very large and highly leveraged gambles and recklessly failed to apply even basic risk-management techniques and controls to these gambles.

Named in the lawsuit were Amaranth's founder, Nicholas M. Maounis; chief operating officer, Charles H. Winkler; energy trader, Brian Hunter; and Robert Jones, the chief risk officer. The lawsuit continues:

... The Net Asset Value (NAV) that defendants reported for the Fund at year-end 2005 included tens of millions of dollars in unrealized "profit" on natural gas positions and spreads without any discount for the illiquidity of those positions. Nonetheless, Advisors paid itself a 20% "performance fee" on this unrealizable "profit" and awarded huge bonuses to those responsible for the operations of the Fund, i.e., Maounis, Winkler, Jones, and Hunter. Hunter alone, upon information and belief, received a $100 million bonus for 2005!

The chief risk officer, Robert Jones, was reported to have been paid a bonus of at least U.S. $5 million for 2005.[13] Little wonder hedge fund risk managers are seen as the *emperors* of financial risk management, rather unfortunately, like Emperor Nero accused of playing his lyre while standing on the summit of the Palatine as flames devoured the city of Rome; the tantalizing possibility of a very large bonus may affect a degree of insouciance toward risk in those charged with overseeing it.[14] Indeed, the observation of *The Post's* Philip Delves Broughton resonates with many[9]:

Some economists believe hedge funds improve the efficiency of the financial markets by introducing competition, new ideas and liquidity. Amaranth's collapse, however, strengthens the sceptics, who share the belief of the great investor Warren Buffett who, after observing the lavish salaries hedge fund managers paid themselves, said these funds were no more than a compensation structure dressed up as an industry.

The Utter and Total Redundancy of Financial Risk Management

The intrinsic worth of the risk management function has also come into question by management scholars. A major tool used in theoretical finance is the Capital Asset Pricing Model (CAPM) (see Appendix). It states that the total risk of a security can be divided into two components. The first, known as specific or unsystematic risk, reflects variation in a stock's return as a result of firm-specific events, such as the development of a new product or the firing of a chief executive officer. The second component of risk, known as systematic or market risk, reflects the variation in a stock's return as a result of economy wide events such as an economic slowdown, or natural disaster. Thus, according to the CAPM model either economy wide or firm-specific events can alter a stock's price.

Numerous empirical studies have shown that for an individual security, unsystematic risk is by far the largest component of total risk.[15] Typically, it is around 50% to 80% of the total risk of an individual stock.[16] However, because unsystematic risk can be diversified away, it is not relevant to a diversified investor. In practice, a portfolio with as few as 15 to 20 stocks may be sufficient to eliminate most of the unsystematic risk.[17] Since unsystematic risk can be diversified away, it will not be reflected in investor's valuation of the firm.[18] The CAPM therefore predicts that well-diversified investors will be primarily concerned about systematic risk. As Bettis (1983) points out, if the firm's objective is to maximize shareholder value, then this prediction:

> … leads naturally to the conclusion that managers should not manage unsystematic risks … because such behavior will not be rewarded by the stock market.

Thus, under the assumptions of the CAPM, academic scholars have postulated that the risk management function, whose primary focus is the management of risk associated with operation of the firm, may be redundant.[19] It is therefore, perhaps, not surprising that one often detects an ambivalent attitude toward it by senior executives. The risk management function is somewhat dismissively considered inutile, interesting only to *rocket scientists* and academics.

The Risk Manager as a "Quivering Dastard"

That the risk manager is characterized in popular culture as little more than a quivering dastard is rather unfair. For sure, a few rotten risk managers have undertaken actions befitting a dastard; but that as a consequence, the risk management function should be lambasted as being of no practical value is wholly and utterly mistaken. While finance theory informs us a well-diversified investor will have eliminated unsystematic risk, it does not claim it should go unmanaged. In practice, senior managers may have a very personal incentive to manage it. First, by doing so, they may reduce the likelihood of bankruptcy and thereby enhance their job security. Second, it is not unreasonable to assume managers who invest significant human capital in their specific businesses will be concerned about the totality of risk. Third, if senior management is compensated on the basis of their firm's earnings, they may prefer a stable to a volatile earnings stream. Fourth, talented managers may actively seek to avoid firms where unsystematic risk, and therefore career risk, is high. To the extent that reduction of unsystematic risk may increase the quality of management, it may be an appropriate consideration for investors. As Salter and Wainhold (1979) explain:

> Given a business opportunity producing a cash flow, the risk/return model emphasizes that market value will be affected by managing systematic risk rather than unsystematic, or company specific risks. Ironically, managers spend most of their efforts on these very real company specific risks. Managers do this because company specific risks (such as competitive retaliation, labor relations, or even bankruptcy) are both obvious and immediate, as well as being potentially disastrous to personal and organizational welfare.[20]

Despite widespread acerbity toward it and considerable academic cogitation about it, there can be little doubt that the risk management function is of some, often considerable, value. That the wisdom of Salter and Wainhold convey fundamental truth, which holds firmly across decades, is made piercingly clear by the little known case of David Deutsch. Deutsch, a quick-eyed pointy headed middle-aged executive, was the chief investment officer of SDCERA. His narrow nose, strangely bulbous head, and dark, almost black hair, gave him, in a rather strange but familiar way the bearing of a 19th century English civil servant; perhaps, in an earlier era he would have found himself serving Queen Victoria in India for the British Empire.

The enormity of Deutsch's cranial cavity appeared such that there could be little doubt about his mental capacity. And indeed, he had graduated from the University of Texas with an MBA in finance and held the Chartered Financial Analyst designation. Yet, despite the magnitude of his mental capacity, he apparently ignored, forgot, or was unaware of the enduring wisdom of Salter and Wainhold. For Deutsch presided over the SDCERA portfolio as the hedge fund Amaranth imploded. He survived, but only just. Unfortunately for Deutsch, a few years later, following the economic collapse of 2008, a $2.5 billion loss in the pension assets he oversaw, and the collapse of yet another hedge fund in his portfolio, he resigned. While one understands the necessity for asserting this was under Deutsch's own free will. It appears he left SDCERA without a severance package.[21] Although little discussed by academics or written about in scholarly journals, the case of David Deutsch serves to highlight the absolute relevance and critical importance of effective risk management. Without it, risk for executive leaders such as chief investment officers is unpalatably high. And those who do not acknowledge this fundamental fact will find themselves, sooner or later, in the unfortunate position of Deutsch—that is, seeking alternative employment.

In practice, even the mordaciously nefarious corporate executive understands the inherent marketing and public relations power of risk management[2]:

> It would be hard to think of a more important concept for Enron than managing risk. ... When reporters and analysts inquired about the company's risk management abilities, Skilling [Enron's Chief Executive Officer] had a ready answer: he pointed to Enron's Risk Assessment and Control department, known inside the company as RAC. ... Skilling knew Wall Street wanted to see a strong system of internal controls and after he was named president, he made RAC a centerpiece of management presentations to Wall Street analysts, investors, and credit-rating agencies. ... [RAC's] mission was to assess the economic, financial, credit and political risk in every Enron deal of more than $500,000. ... Most of all ... as Enron described it, RAC had independence and clout. ... Thanks to RAC, Enron was able to portray itself as a company that could safely take on more risk than other companies, precisely because it had the right controls in place. ... Wall Street was dazzled.

So complete was Enron's deception that it was lauded for its sophisticated financial risk management tools and processes. In January 2000, *Risk* magazine named it the "Energy/Commodity Derivatives House of the Year" and in 1999, Andrew Fastow, Enron's CFO, received *CFO* magazine's award for Excellence in Capital Structure Management.[21] W. Chan Kim and Renée Mauborgne writing in 1999 for the *Financial Times* gushed lovingly[22]:

> For four years running, *Fortune* magazine has ranked Enron, the Houston-based energy company that operates in two of the oldest industries in the world—gas and electricity—as the most innovative company in the U.S. Today Enron has as many traders, analysts, and rocket scientists—including a genuine ex-rocket scientist from the former Soviet Union. ... Enron exemplifies the transition from the production to the knowledge economy ...

Enron's use of risk management as primarily a sales and marketing asset is rather unfortunately not that unusual. Symbolic of the esteem with which Amaranth was held in the industry, a matter of months before the ill-fated natural gas trades which forced it to close, it was short-listed for the MARHedge[24] multistrategy hedge fund Performance Awards.[25] Sophisticated risk management also appeared to be a key value proposition in its sales and marketing pitch—The San Diego County Employees Retirement Association (SDCERA) lawsuit states[13]:

> ... In March 2005, Maounis, Winkler, and Jones touted Advisors's supposedly rigorous risk control programs to SDCERA and its consultant during SDCERA's due diligence visit. ... But for these representations, and

the promises of security and sophisticated risk management conveyed to SDCERA, SDCERA would not have invested in the Fund. Having fraudulently induced SDCERA's investment, defendants' fraud and misrepresentation continued throughout the duration of SDCERA's participation in the Fund. SDCERA was repeatedly assured by defendants that the Fund's exposure to volatile energy markets was being reduced and hedged. ... However, in truth, defendants were not reducing or hedging against the Fund's natural gas exposure, and, instead, were recklessly deepening the Fund's positions in natural gas and disregarding basic risk management principles, actions that eventually caused the collapse of the fund. ... Beginning in 2005, the Fund, however, was being run, either intentionally or negligently, as a de facto single-strategy natural gas fund, placing billions of dollars at risk in highly volatile markets and with no exit strategy. Advisors encouraged Hunter to take such enormous, leveraged positions in natural gas markets that the Fund's own trading volume influenced spreads and prices in these markets.

Perception and Reality about Risk Management

The process of financial deregulation, disintermediation, and globalization alongside periodic financial crisis has led to increased pressure on an institution's risk-taking practices; and this, in turn, has heightened the need for effective risk management. As Vincent Oliva, then vice president and research director for Gartner, observed[26]:

> The elements of risk management have always been a concern for financial institutions, but during the past few years risk and risk management have become even more serious and complex issues and now have much greater visibility and priority within the organization. ... Since ineffective risk management can have a serious negative impact on a financial institution's bottom line, it is now a critical, all-encompassing concern for the financial services industry.

In 1993, the Group of Thirty[27] recommended that market and credit risk management should be independent functions of the firm. This provided a powerful incentive for multinational banks to take corporate-wide risk management seriously. Two decades later, the objective, function, and value of corporate-wide risk management remains far from settled. Yet, while academics are still debating many of the issues, practitioners are exploiting risk management for competitive gain or personal enrichment. Companies such as Enron and Amaranth clearly understood and exploited the marketing and sales power of risk management. Yet, they failed to grasp the fact that rigorous and continuous management of risk lies at the heart of effective business strategy. Strong risk management frees the firm to exploit good risks while limiting downside exposure.

Since most financial decisions, whether they concern capital structure, dividends, capital allocation, capital budgeting, or investment and hedging activity, revolve around the trade-off between taking on a risk and the expected rewards, how effectively a firm determines risk versus reward will determine how efficiently they leverage their resources.

To reap the benefits, senior management must design, support, and engage actively with their risk management function. They must explicitly acknowledge and sincerely believe the management of risk is a determinant of how their firm will grow and whether it will flourish or decline and ultimately die. The notion of risk management must be firmly embedded in corporate culture if the true benefits are to be reaped. Effective risk management is a valuable asset to the firm. From a regulatory and ratings perspective, robust risk management, enforced by a regulator and monitored by a ratings agency, may ultimately lead to a lower cost of capital.

Where risk management is weak or inadequate, we should not be surprised when calamity eventually follows. That dramatic collapses such as IndyMac,[28] Enron, Amaranth, or Lehman Brothers occur is not an argument against risk management per se, but raises the question as to whether there was enough risk management, or the right risk management in place. The failure of boards or senior managers to firmly establish and enforce adherence to effective risk practices, compensation programs that conflict with the underlying ethos of risk management, and command and control structures that disallow independence in the risk management function all serve to undermine the value added of risk management.

Yet, risk management is clearly an input into the firm's value creation processes, just as labor, capital, land, and entrepreneurship are the traditional economic inputs into a firm's production function. It is imperative, therefore, to get risk management right. Firms that do so enhance their competitive edge.[29] Value-added risk management is the process of aligning an organization's risk management activity with strategic business objectives. It is about incorporating sound risk management practices into the decision-making process, thereby better equipping senior management with the ability to ask informed questions and make more rational decisions. At its core is the desire to provide senior management with better knowledge about risk in order to enhance return and sustain a competitive advantage. Unfortunately, we do not yet live in a world where the science of risk management is fully established with risk professionals taking their place alongside lawyers and accountants. Risk management textbooks do not prepare the risk practitioner with sufficient empirical and theoretical ammunition to counter the perception that the risk management function is a total and utter waste of time. And this is a real shame because risk management matters! And this remains for many, the unpalatable truth about risk management.

KEY POINTS

Key Point 1: On the trading floor and in executive offices the risk management function is frequently characterized as a bureaucratic destroyer of value, an unnecessary consumer of scarce corporate resources and a total and utter waste of time.

Key Point 2: The situation has not been helped by the widespread characterization of the risk manager as an evil dastard. Rather unfortunately, a handful of sycophantic risk managers who lacked moral fiber have undertaken actions befitting a dastard.

Key Point 3: Academic scholars have added to the debate by pointing out that the CAPM predicts well-diversified investors will be primarily concerned about systematic risk. If the firm's objective is to maximize shareholder value then this prediction leads to the conclusion that managers should not manage unsystematic risks because it will not be rewarded by the stock market. Academic scholars have postulated that this may imply the risk management function is redundant.

Key Point 4: Despite widespread processional acerbity toward it and considerable academic cogitation about it, there can be little doubt that risk management is of value. Senior managers have a strong incentive to manage risk.

Key Point 5: The failure of corporations such as Enron and Lehman Brothers is not an argument against risk management but a reason for its effective enforcement.

Key Point 6: However, risk management functions, which are neither properly resourced nor enforced, will be effective.

Key Point 7: Weak and ineffective risk management represents a failure in senior management's vision rather than an inherent weakness in the fundamental concept of risk management.

Key Point 8: The benefits of a risk management function can only be reaped if senior management engages actively with it.

Key Point 9: Risk management is a determinant of how the firm evolves—whether it will grow and flourish or flounder and perish.

Key Point 10: Continuous management of risk lies at the heart of effective corporate strategy. It frees the firm to exploit good risks while limiting downside exposure.

Key Point 11: Integrity, intelligence, and industry are essential qualities of a risk manager. Those charged with heading the risk management function should be neither fearful nor willing to grant *favors* to anyone.

Key Point 12: Value-added risk management is founded on the desire to provide senior management with better knowledge about risk in order to enhance return and hence competitive edge. At its core, it is about:

- Aligning an organization's risk management activity with strategic business objectives.
- Incorporating sound risk management into the decision-making process.
- Better equipping senior management with the ability to ask informed questions and make rational decisions.

For Further Thought

This chapter has focused on perceptions about and the reality of risk management. The place of the risk management function in the corporate hierarchy remains an issue of considerable debate. The discipline itself has struggled to grace itself with the professional aura one associates with corporate lawyers or accountants. Part of the problem may lie in the observation that it is not a designated career destination but rather lies between the cracks of compliance and revenue generation. It is an area one falls into *on the way to the trading floor* or else, following failure, *on the way back from the trading floor*. Yet, the management of risk is an important element in business strategy and the tactical implementation of that strategy. Robust, reliable, and accurate risk management, measurement, and assessment are critical to business success.

Issues to consider when exploring the perceptions and reality of risk management include:

1. What is the place of risk management and the risk management function in your organization?
2. What is the level of power and influence delegated to the function?
3. To whom is it accountable? Does that individual hold executive office?
4. Is risk management seen as a career objective, final goal, or a stepping stone to other activities?
   - i. Where do your risk managers come from?
   - ii. How are they compensated?
   - iii. How long do they stay?
   - iv. Why do they leave?
   - v. Where do they go?
5. What is the level of seniority and respect shown to risk managers?

6. Is risk management serving some bureaucratic corporate function or adding valuable input into the decision-making process?
    i.   How do you know?
    ii. How is the *value-added* of the function assessed?
7. What is the mechanism by which risk management analysis is assimilated into the decision-making process?

Additional Resources

This chapter has touched upon a number of issues that are important for the individual firm and the orderly functioning of a modern economy. Bettis (1983) and Peavy (1984) explore the disconnect between modern financial theory's irrelevance view of risk management and the value-added perspective of corporate strategy theory. King (1966) provides insight into risk from the perspective of systematic and industry-specific factors. Lewis (2003, 2004, 2005, 2008a, 2008b, 2008c, 2009a, 2009b, 2009c), Lewis and Okunev (2008, 2009) and Lewis, Okunev, and White (2007) outline how statistical and quantitative procedures can be of assistance in identifying, measuring, and modeling risk. McLean and Elkind (2003) illustrate the disastrous consequence of feeble and ineffective risk management. The Group of Thirty (1993) and Tippins (2004) provide additional insight into the role of risk management and where it belongs in the corporate hierarchy. Reinhart and Rogoff (2009) discuss the role of systematic risk primarily within the context of financial crises. The original irrelevance of corporate risk management argument is addressed in the classic work of Modigliani and Miller (1958).

Bettis, K.A. (1983). Modern Financial Theory, Corporate Strategy, and Public Policy: Three Conundrums. *Academy of Management Review* 8:406–414.

Group of Thirty. (1993). *Special Report on Global Derivatives. Derivatives: Practices & Principles*. Washington, DC: Group of Thirty.

King, B.F. (1966). Market and Industry Factors in Stock Price Behavior. *Journal of Business* (I):139–190.

Lewis, N.D. (2003). *Market Risk Modelling: Applied Statistical Methods for Practitioners*. London: Risk Books.

———. (2004). *Operational Risk with Excel and VBA*. New York: John Wiley.

———. (2005). *Energy Risk Modeling: Applied Modeling Methods for Risk Managers*. London: Palgrave.

———. (2008a). Making Ends Meet: Target Date Investment Funds and Retirement Wealth Creation. *Pensions: An International Journal* 13(3):130–135.

———. (2008b). The Relationship between Target Date and Target Risk Funds. *Pensions: An International Journal* 13 (1 and 2):55–60.

———. (2008c). Assessing Shortfall Risk in Life-Cycle Investment Funds. *Journal of Wealth Management* 11(1): 15–19.

————. (2009a). Assessing the Impact of Fees on Performance and Shortfall Risk in Target Date Investment Funds. *Journal of Investing* 18(4):72–78.

————. (2009b). Is There a Role for Commodities in Long-Term Wealth Accumulation? *Journal of Wealth Management* 12(2):130–137.

————. (2009c). Using Hedge Funds to Enhance Asset Allocation in Life-Cycle Pension Funds. *Pensions: An International Journal* 14(1):52–55.

Lewis, N.D. and Okunev, J. (2008). Estimating the Risk of Guaranteed Products. *Journal of Investing* 17(3):86–95.

————. (2009). Using Value at Risk to Enhance Asset Allocation in Life-Cycle Investment Funds. *Journal of Investing* 18(1):87–91.

Lewis, N.D., Okunev, J., and White, D. (2007). Using a Value at Risk Approach to Enhance Tactical Asset Allocation. *Journal of Investing* 16.

McLean, B. and Elkind, P. (2003). *The Smartest Guys in the Room*. New York: Portfolio Books.

Modigliani, F. and Miller, M. (1958). The Cost of Capital, Corporation Finance, and the Theory of Investment. *American Economic Review* 48.

Peavy, I.W. (1984). Modern Financial Theory, Corporate Strategy, and Public Policy: Another Perspective. *Academy of Management Review* 9:152–157.

Reinhart, M.C. and Rogoff, S.K. (2009). *This Time Is Different: Eight Centuries of Financial Folly*. Princeton, NJ: Princeton University Press.

Tippins, S.C. (2004). Risk Management: Where Is It and Where Does It Belong? *Risk Management: An International Journal* 6(3):35–50.

Appendix

The Capital Asset Pricing Model is a standard tool used in both theoretical and applied finance. We detail some of its important characteristics in the box below.

THE CAPITAL ASSET PRICING MODEL

For a diversified investor, CAPM states that the required rate of return, say, on IBM stock is proportional to its systematic risk measured by beta (β):

$$R_{IBM} = R_f + \beta_{IBM}(R_M - R_f),$$

where

(i) R_{IBM} is the rate of return on IBM stock,

(ii) R_f is the risk-free rate of return, that is, return on government bills/bonds,

(iii) and R_M is the overall market rate of return such as the FTSE Allshare index in the UK or the Russell 3000 index in the U.S.

β_{IBM} captures the sensitivity of the IBM's return to movements in the overall market and is therefore a measure of systematic risk. In the above example, β_{IBM} is related to IBM's price risk measured by volatility (σ_{IBM}), multiplied by its correlation with the overall market ($\rho_{IBM,M}$), and divided by the overall market risk (σ_M):

$$\beta_{IBM} = \frac{\rho_{IBM,M} \times \sigma_{IBM}}{\sigma_M}$$

Furthermore, note that:
  (i) The beta of the overall market is 1.
  (ii) If β_{IBM} is less than 1, then IBM's stock has less systematic risk than the market as a whole.
  (iii) If β_{IBM} is greater than 1, the stock will have more systematic risk than the market as a whole.

Endnotes

1. MBA or Master of Business Administration is a degree in business administration, which serves primarily as a gateway into leading corporations.
2. See McLean and Elkind (2003), Chapter 9, pp. 114–117.
3. See Lewis, N.D. (2005).
4. Which claimed mental illness was a result of infectious material residing in the roots of teeth and recesses in the bowels.
5. See Mitchell, Donna, (2006), Doubts Plague Risk Management, *Asset Securitization Report* 6(14), April 10.
6. See Philip M Halperin (http://www.btinternet.com/~phalperin/).
7. See Tippins (2004).
8. See, for example, the documentary, *Enron: The Smartest Guys in the Room* (2005), directed by Alex Gibney.
9. See *The First Post*, an independent daily news magazine (http://www.thefirst-post.co.uk/).
10. The relevant passage reads:
   "Immortal amarant, a flower which once
   In Paradise, fast by the tree of life,
   Began to bloom; but soon for man's offence
   To Heaven removed, where first it grew, there grows,
   And flowers aloft shading the fount of life,
   And where the river of bliss through midst of Heaven
   Rolls o'er Elysian flowers her amber stream;
   With these that never fade the Spirits elect
   Bind their resplendent locks inwreathed with beams.
   (John Milton, 1667, *Paradise Lost*, Book III)

11. The typical life span of a hedge fund is between three to six years. See Stanford Graduate School of Business (http://www.gsb.stanford.edu/news/headlines/vanhorne_hedgefunds.shtml or http://hedgefundcenter.com/hfc/).

12. The case is *San Diego County Employees Retirement Association v. Nicholas Maounis, Charles Winkler, Robert Jones, Brian Hunter and Amaranth Advisors LLC*, U.S. District Court, Southern District of New York (Manhattan). The case was dismissed by the court on March 18, 2010.

13. See *The Wall Street Journal* online. (For further details see: http://online.wsj.com/article/SB117012628156391989-search.html.)

14. Some large hedge funds are diligent in policing risk, but many managers have small workforces and focus on trading. The sheer size of potential payouts to senior risk managers leads one to question their ability to remain an impartial and independent overseer of risk—more concentrated positions increase both risk and bonus potential! This circumstance not only attracts to the position scoundrels, but also turns people who should know better into scoundrels.

15. See, for an early example, King (1966).

16. See King (1966), Bettis (1983), Peavy (1984), and Tippins (2004).

17. See, for example, Bettis (1983).

18. The diversification concept can be more easily understood if one considers an investor who holds a portfolio that exactly replicates a selected market indicator such as FTSE All share index. Since this portfolio is exactly equivalent to the market, it must consist primarily of systematic or market risk; in other words, unsystematic risk has been diversified away.

19. In the stylized Modigliani and Miller (1958) world of corporate finance theory, neither capital structure nor corporate risk management affects the value of the firm. Since investors can diversify away firm-specific risk, they will not be rewarded for taking it on. The issue of whether the management of unsystematic risk is compatible with modern financial theory is debated in, for example, Bettis (1983) and Peavy (1984).

20. See Malcolm Salter and Wolf Wainhold, *Diversification through Acquisition*, Free Press, New York, 1979.

21. See the *San Diego Union Tribune* (http://www.signonsandiego.com/).

22. See Rosen, Robert Eli, (2003), Risk Management and Corporate Governance: The Case of Enron, *Connecticut Law Review* 35(3).

23. See W. Chan Kim and Renée Mauborgne, (1999), New Dynamics of Strategy in the Knowledge Economy, *The Financial Times*, FT Mastering Series, November 10.

24. Held in April 2006. MARHedge is a hedge fund industry information service offering editorial bureaus in key financial markets, publications, news gathering, and dissemination as well as organizing global conferences business.

25. It eventually lost out to Ore Hill Partners LLC, who was awarded the MARHedge Multistrategy Hedge Fund Performance award.

26. Comments reported April 23, 2003. See *Information Management* online (http://www.information-management.com).

27. A private, nonprofit international body composed of senior representatives of the private and public sectors and also academia. See Group of Thirty (1993).

28. The actual loss was absorbed by the Federal Deposit Insurance Corporation. See Federal Deposit Insurance Corporation (http://www.fdic.gov/bank/individual/failed/IndyMac.html) for further details.

29. By reducing the likelihood of bankruptcy, diminishing agency problems, and ultimately lowering the cost of capital.

Section II

What You Need to Know, But Nobody Wants to Tell You

We turn now to some observations that are often overlooked by those involved in the management of risk. We begin by tracing out the link between corporate governance and risk management. Next, in Chapters 5 and 6 we will consider the core lessons surrounding the role of risk managers. We touch on the difficult subject of integrated, single lens analysis of risk in Chapter 7. The subsequent two chapters cover areas risk managers should think more deeply about, but rarely do. We will end in Chapter 10 with a discussion about that most dominant of risk measures—value at risk.

Section II

What You Need to Know, But Nobody Wants to Tell You

4

What the Textbooks Will Not Tell You about Corporate Governance

When the Mighty Sparrow[1] first penned the words of the calypso "Drunk and Disorderly," little did he know it would take the island of Trinidad by storm, sweeping both the 1972 Carnival Road March and Calypso Monarch titles into his arms. The theme resonated with the West Indian diaspora, and it went on to become one of the greatest calypso songs of all time. In an economic environment still reeling from decades of unsustainable growth in personal debt, economic distress following the 2008 financial crisis, persistently high unemployment, revelations of *rock star* style executive excess and massive levels of government spending, the key verse of that song, repeated over and over in its many classical, steel-pan, and soca[2] incarnations, strikes a consonant chord in those concerned about the dismal state of corporate governance in the Western world:

> Drunk and disorderly
> Always in custody
> Me friends and me family
> All man fed up with me, cause I
> Drunk and disorderly
> Every weekend I in the jail
> Drunk and disorderly, nobody to stand me bail

In the late 1990s and early 2000s, corporate wrongdoing dominated newsprint and evening news television programs. Right in the heart of American capitalism, before the eyes of a stunned world, emerged a band of drunken executive tossers, modern day *robber barons* who self-righteously debauched their firms, destroying investor value and in the process ruining the retirement dreams of their hardworking employees. Malfeasance and fraud were rampant in companies like WorldCom, which hid U.S. $3.8 billion in expenses; Enron, which hid U.S. $1 billion in debt; and Tyco, whose chief executive officer (CEO) was indicted for evading U.S. $1 million in sales tax. So vulgar and villainous were the stories emerging out of corporate America in 2002 that the normally genteel *Seattle Times* observed with caustic exasperation[3]:

> Wall Street's 2002 will be remembered as a parade of catastrophe and disgrace. It will be hard to forget the image of executives of billion-dollar

companies led, handcuffed, to the courthouse steps. ... Amid a stock-market plunge and an eight-year high in U.S. unemployment, the avarice exposed in 2002 was astonishing. ... As details emerged about Enron, WorldCom, and executives who got rich at the expense of employees and shareholders, stocks plunged, jobs were lost and retirement savings wiped out.

Almost a decade later, and in the wake of the worst financial conditions since the Great Depression, reports began to emerge of how investment firms on Wall Street had actively contributed to the financial meltdown of 2008. It appeared leading investment houses benefited from what they saw as the coming housing collapse at the same time as actively encouraging lowered lending standards. In the autumn of 2008, the credit markets across the world froze and liquidity evaporated. Foreclosure filings in the United States hit a record high in the third quarter of 2009. It was the worst three months for housing in United States economic history. In that quarter alone 937,840 homes received a foreclosure notice—around one in every 136 U.S. homes: there were almost 4 million foreclosure filings during 2009.[4] Mortgage giants Countrywide and IndyMac were prominent casualties of the collapse in house prices.

While it is certainly true that an Englishman's house is his castle, it is equally true that home ownership had become an inherent part of the American dream; a dream, which by 2009, had metamorphosed into a horrendous nightmare for millions of hardworking families. As with all major calamities, people began to point fingers, initially at the politicians, who, in turn, pointed at Wall Street. And Wall Street complied by offering up a bounty of truly stomach-wrenching scandals. Bernard Madoff's breathtaking $65 billion Ponzi scheme shook many investors' faith in the integrity of the financial system. Madoff was sentenced to 150 years in prison.

During the same year, the global Swiss bank UBS handed over $780 million to settle criminal charges that it helped American citizens evade taxes using concealed offshore accounts. The flamboyant businessman Tom Petters was found guilty of running a $3.6 billon dollar Ponzi scheme. It was the largest financial fraud in the history of the state of Minnesota and cast a long ugly shadow over the quaint midwestern town of Minnetonka.[5] Raj Rajaratnam, renowned for his hedge fund, Galleon Group, was arrested by the FBI on allegations of insider trading.[6] Robert Allen Stanford, the first American to be knighted by the tiny islands of Antigua and Barbuda and celebrated for his financial empire, was charged with fraud.[7] On and on the financial scandals continued to emerge with many of their victims, too numerous to mention, cast unceremoniously into penury and want. As 2009 slowly progressed, public anger began to grow louder. The impassioned words of Ann Minch of Red Buff, California captured the *zeitgeist*, "You are evil, thieving bastards...Stick that in your bailout pipe and smoke it."[8]

By mid-April 2010, so vicious was the public mood that President Obama, in a highly symbolic move, traveled to the heart of the U.S. financial markets—downtown Manhattan. There, he demanded financial regulation and legislative reform. Within a week or so,[9] thousands of workers poured onto Wall Street chanting "Bust up! Big banks!" and "People power!" The throng stormed the offices of JPMorgan Chase and Wells Fargo. It appeared to be an instantaneous outburst of public anger fueled by a massive loss of jobs, the size of government bailouts to financial institutions, huge and growing government deficits, and perhaps most symbolic of all, anger at a 140-year-old financial institution by the name of Goldman Sachs.

On the 16th of April 2010, in a quite astonishing turn of events, the United States Securities & Exchange Commission (SEC) filed securities fraud charges against Goldman Sachs and a former employee by the name of Fabrice Tourre.[10]

> The Commission brings this securities fraud action against Goldman, Sachs & Co. ("GS&Co") and a GS&Co employee, Fabrice Tourre ("Tourre"), for making materially misleading statements and omissions in connection with a synthetic collateralized debt obligation ("CDO") GS&Co structured and marketed to investors. This synthetic CDO, ABACUS 2007-ACI, was tied to the performance of subprime residential mortgage-backed securities ("RMBS") and was structured and marketed by GS&Co in early 2007 when the United States housing market and related securities were beginning to show signs of distress. Synthetic CDOs like ABACUS 2007-ACI contributed to the recent financial crisis by magnifying losses associated with the downturn in the United States housing market.

U.S. Senator Carl Levin, a Michigan Democrat, who led the Senate's powerful Permanent Subcommittee on Investigations, posted internal e-mails on his Web site that he said showed the investment bank Goldman Sachs "made a lot of money by betting against the mortgage market." And in a quite shocking press release it was claimed[11]:

> Investment banks such as Goldman Sachs were not simply market-makers, they were self-interested promoters of risky and complicated financial schemes that helped trigger the crisis," said Sen. Levin. "They bundled toxic mortgages into complex financial instruments, got the credit rating agencies to label them as AAA securities, and sold them to investors, magnifying and spreading risk throughout the financial system, and all too often betting against the instruments they sold and profiting at the expense of their clients." The 2009 Goldman Sachs annual report stated that the firm "did not generate enormous net revenues by betting against residential related products." Levin said, "These e-mails show that, in fact, Goldman made a lot of money by betting against the mortgage market.

Over in the United Kingdom the situation was no better. In May 2010, the governing Labour Party, which had risen to power in 1997 under the leadership of the charismatic politician Tony Blair, was tossed out of office. The so-called moat-gate scandal ignited a public backlash against all politicians. At a time when millions of workers were losing their jobs and thousands of homes were being repossessed, it was discovered that Douglas Hogg, a former cabinet minister, included with his political expenses the cost of having his moat cleared, piano tuned, and stable lights fixed at his ritzy country manor house. He was not alone in expecting the taxpayers to subsidize his exclusive lifestyle. Politicians from all sides of the spectrum were caught with their *fingers in the public till*. Revelations such as these drove a stake into the governing administration's chance of reelection[12] and cemented distrust of all politicians firmly in the mindset of the British public. The outcome of the May 2010 elections was a hung parliament with no one political party able to form a government, a result welcomed by the British people.

By the end of the first decade of the 21st century, it seemed clear to all that recreant corporate officers and ignoble politicians knew no bounds. It seemed, as with the drunk in the Mighty Sparrow calypso, aberrant behavior was compulsive and they would not, could not stop:

> I can't stop Lord and I won't try,
> I feel so good when I high
> So bring wine, bring beer, bring gin,
> Bring champagne, ink anyting
> Night time, better in the sunshine,
> Anytime is right time,
> To drink me high wine, until I get
> Drunk and disorderly ...

The Essence of the Governance Issue

Confidence in the capital markets rests on a complex network of supporting institutions that ensure shareholders receive trustworthy information about the value of a company and give assurance that the managers and the controlling shareholders will not cozen smaller investors out of most or all of the value of their investment. In an ideal world, we would expect corporate executives and senior managers to act ethically, even altruistically. This, however, is not the message received from the scandals of recent years. Republican Senator John McCain, former presidential candidate, reflecting on the Goldman Sachs issue exclaimed[13]:

> I don't know if Goldman Sachs has done anything illegal ... there's no doubt their behavior was unethical, and the American people will render a judgment as well as the courts.

The dramatic collapse of Lehman Brothers, Countrywide, Enron, WorldCom, Parmalat,[14] and the uproar surrounding Goldman Sachs naturally casts doubt on the efficiency and effectiveness of corporate governance practices and their enforcement by the regulators. In each and every reported corporate crisis, there appears to have been a failure in part or wholly of the board, the auditors, or market mechanisms such as investment banks and rating agencies, which might have been expected to give a warning of trouble ahead.[15]

The key governance problem is the separation of ownership and control. By law, shareholders own the company, and managers' fiduciary duty is to work on behalf of shareholders to allocate business resources to their optimum use. In practice, for companies listed on stock exchanges, de facto shareholder control is generally weak with few shareholders having sufficient holdings to individually influence the choice of boards of directors or chief executive officers. Shareholders generally exercise little control over either day-to-day operations or long-term policy. Instead, control is vested in the hands of professional managers. As Berle and Means famously explained[16]:

> The separation of ownership from control produces a condition where the interests of owner and of ultimate manager may, and often do, diverge ...

Preventing such divergences, or at least minimizing their effects is the role of reputational intermediaries such as investment banks, audit firms, and regulators such as the Securities & Exchange Commission in the United States and the Financial Services Authority and Financial Reporting Council in the United Kingdom. It is the responsibility of a firm's senior management to organize and control its activities responsibly with adequate risk management systems and controls. Good corporate governance is therefore best understood as a well-functioning system of corporate direction, regulatory oversight, and control through the market mechanism.

The Superficiality of Compliance

If anybody can understand the miserable plight of many a risk manager, it is perhaps the compliance officer. For they themselves are knowing parties to an impecunious discipline whose perpetuity is secured by dint of legislation and regulation. As is underscored in my oft-recounted tale, this knowledge affords the compliance officer a degree of self-aggrandizement, which is rarely evident in even the highest-ranking risk manager.

> It was Christmas, and I found myself standing in the corner of yet another male-dominated office party, this time, operations. Talking to a rather bland accountant about nothing in particular, I noticed, out of the corner of my eye, a wiry little man bouncing high on the tips of his toes around the edge of the room. His head was shaven and he had a heavy gold chain wrapped loosely around his neck. He was not wearing the usual attire of shirt and tie but rather, a crumpled grey cotton jacket with black turtleneck top and diamond cut carpenter blue jeans; on his feet he wore sneakers, Addidas *All Black*. He swaggered around the room like a CEO with his pearly white teeth glistening as he spoke. "Who is that?" I asked the accountant.
>
> "Oh, you mean Andrew, he is our new chief compliance officer, up from California. We were lucky to get him!"

Compliance officers can afford to swagger. Maintaining a corporate compliance program is an unavoidable prerequisite of doing business. It signals to investors, regulators, and legislators a corporate-wide commitment to creating and maintaining a culture that respects and adheres to applicable laws and regulations.

That compliance be associated with good corporate governance is perhaps not too surprising when one considers the definition of corporate governance: a control system designed to establish and maintain internal mechanisms intended to prevent and detect unethical, illegal, and/or poor business practices and violations. Many large firms have appointed a chief compliance officer with responsibility to oversee, monitor, and correct practices. Typical mechanisms of implementation include written policies, procedures, and operational practices.

For many boards and corporate executives, clarity on governance may be clouded by the urgent necessity of satisfying regulatory requirements; and they make the mistake of equating good governance with meeting minimally the demands of regulators and legislators through the corporate office of compliance. That is, they tend to look at governance solely as a compliance exercise[17] with senior executives naturally turning to chief compliance officers to provide assurance. As new regulatory mandates come into force, the compliance office shifts to reactive mode pursuing solutions specifically designed to address individual legislative requirements with no reference to the wider strategic or ethical context.

While a reactive approach to compliance may appear to offer a short-term, *easy* fix, it is a more questionable long-term strategy. This is because it leaves organizations managing a series of possibly unrelated policies and rules aimed at addressing specific regulatory requirements without a common corporate-inspired theme. As Susan Bies highlighted when she was a member of the Board of Governors of the Federal Reserve System[18]:

> ... any institution that views corporate governance as merely a compliance exercise is missing the mark. We all are aware of companies in various industries who have successfully presented their strategic vision to investors but later stumbled because the execution of that strategy did not meet expectations. Although shortfalls can occur for many reasons, one of the more common shortcomings is focusing the strategy itself too much on market and financial results without giving adequate attention to the infrastructure necessary to support and sustain the strategy.

The ultimate consequence of this disconnect is that compliance departments become overly bureaucratic. In the worst case, there arises an emphasis on creating and ticking boxes rather than doing anything meaningful or productive. As compliance officers lose sight of the fact that the business of business is business, they begin to consume the vital resources of senior management and other professionals. Senior management, in turn, find themselves devoting voluminous amounts of time to satisfying the demands of the compliance office. It can ultimately become a vicious spiral downwards, destroying innovation, value creation, and competitive edge. Little wonder, therefore, that in many environments, executive support for compliance is minimal and implementation is at best halfhearted with poorly paid compliance officers focusing on those aspects which are easy to implement because they are cheap, or because everyone else is implementing them, or because they are forced to implement them as a result of regulator action or stakeholder pressure.

That compliance officers are often regarded, along with their auditor colleagues, as little more than aggressive wide-jawed Gnatbobdellida leeches sucking the entrepreneurial blood out of corporations, or else maggot-like night crawlers cast out into the rivers of successful moneymaking businesses in the hope of snagging a large unsuspecting catch is rather a shame. Yet, alas, it is a view held by many. Take, for example, the situation that arose in Australia. The Australian Competition and Consumer Commission (ACCC) has longed pushed businesses to implement consumer protection compliance programs to satisfy the 1974 Trade Practices Act. Compliance with the Act is a cost and widely perceived by businesses to bring with it little shareholder value. As a result, companies pursued a superficial and largely symbolic interpretation[19]:

> Our survey results on the extent of implementation of trade practices compliance systems by Australian businesses certainly show that implementation is overwhelmingly partial and possibly symbolic. Most businesses have implemented some, but far from all, of the compliance system elements considered by the ACCC, practitioners, and scholars to be necessary for effective compliance management.

Thus, we see a degree of superficiality about the way in which governance practices are implemented and monitored; and with it a real danger that compliance becomes little more than a smoke screen behind which corrupt practices can continue unabated. It is important to realize that putting in place a compliance program is a necessary but not a sufficient condition for good governance. One should not therefore take too much comfort from the swagger of the chief compliance officer.

Why "Gentleman's" Agreements Do Not Work

Attempts at encouraging the practice of good corporate governance in countries such as the United Kingdom and United States are not new. Forward thinking publicly traded, privately held institutions and regulators in these countries have a long tradition of taking governance issues seriously. In the United Kingdom, a drive toward better governance came in the late 1980s in the wake of corporate abuses, including the theft of assets as in the case of Barlow Clowes,[20] the misuse of pension funds in the case of Maxwell,[21] and share price manipulation by Guinness' directors.[22] However, in the typically genteel way of the British aristocracy, the initial response came not in terms of more prescriptive legislation, but rather, as noted by Howard Davies, the chairman of the Financial Services Authority at the time, of self-regulation and best practice guidelines[23]:

> Over the last 15 years, as concerns about corporate governance have grown, a series of codes of practice have been put together, largely by British companies themselves. The basic corporate governance code was designed by Sir Adrian Cadbury. Since then it has been supplemented by another Greenbury code on disclosure of pay, and consolidated, together with a number of other requirements, into the combined code, which is known as the Hampel Code.

The view inherent in these codes of conduct was that the system of British corporate governance was basically sound—as the Hampel Code made clear, there was little need for major overhaul or prescriptive government legislation[24]:

> Public companies [in the United Kingdom] are now among the most accountable organizations in society. They publish trading results and audited accounts; and they are required to disclose much information about their operations, relationships, and remuneration and governance arrangements. We strongly endorse this accountability and we recognize the contribution to it made by the Cadbury and Greenbury committees. But the emphasis on accountability has tended to obscure a board's first responsibility—to enhance the prosperity of the business over time. Business prosperity cannot be commanded. People, teamwork,

leadership, enterprise, experience, and skills are what really produce prosperity. There is no single formula to weld these together, and it is dangerous to encourage the belief that rules and regulations about structure will deliver success.

The perception of a well-functioning corporate governance system "in our country" is not unique to the United Kingdom. For example, Eric Mayne, the then chairman of the Australian Securities Exchange Corporate Governance Council and chief supervision officer for the Australian Securities Exchange (ASX) explained at the launch in 2007 of their revised code:

> This is the first revision of the Council's corporate governance Principles since they were issued in March 2003. This is testimony to the durability of Australia's flexible, principles-based approach to corporate governance. ... There are no drastic or wholesale changes to the Principles. The enduring workability of Australia's governance framework has allowed the Council to fine tune its approach rather than undertake a rewrite. ... Overall, [there was] strong support for the [Corporate Governance] Principles and the "if not, why not" approach to corporate governance disclosure.[25]

The *if not, why not* (also known as *comply or explain*) approach to corporate governance requires companies to either describe how they comply with each of the recommendations that make up a code of practice or else explain why they have chosen not to comply. The approach has been adopted in a diverse range of countries including Cyprus, Norway, the Ljubljana Stock Exchange in Slovenia, and the Stock Exchange of Thailand. These and many more countries have aligned themselves firmly with the spirit of Hampel.

The desire to avoid prescriptive legislation because "it is dangerous to encourage the belief that rules and regulations about structure will deliver success," naturally extends itself to enforcement of codes of practice. The Australians have, perhaps, mastered this subject better than anybody else[26]:

> The ASX [Australian Securities Exchange] Corporate Governance Council's recommendations are not mandatory and cannot, in themselves, prevent corporate failure or poor corporate decision making. They are intended to provide a reference point for companies about their corporate governance structures and practices.

However, such a strong emphasis on the voluntary adoption of corporate governance recommendations naturally leads one to ponder the mechanism by which the governance code is expected to lead to an improvement in governance practice. This concern is particularly apt in the case of Australia, where at the time of the release of their revised governance code in 2007, the term *best practice* was removed altogether from their guidelines:

"Best practice" has been removed from the title and the text of the document—to be known as the ASX Corporate Governance Council's Corporate Governance Principles and Recommendations—to eliminate any perception that the Principles are prescriptive and so not to discourage companies from adopting alternative practices and "if not, why not" reporting where appropriate. (Australian Securities Exchange Corporate Governance Council press release, August 2007)

If governance codes are not *best practice* what are they? Consider for a moment the situation at Enron. How would a voluntary code such as ASX Governance Principles and Recommendations have altered the course of the ensuing calamity? Judging by the attitude of Enron's senior management, probably not by very much:

I have gone back and tried to think what I would have done differently given the facts at the time. And quite frankly, there is nothing I can come up with. (Jeff Skilling, president and chief operating officer of Enron)[27]

The real issue is what lessons can be learnt from past corporate governance failures? As Daniel Henninger of *The Wall Street Journal* notes[28]:

When a company called Enron … ascends to the number seven spot on the Fortune 500 and then collapses in weeks into a smoking ruin, its stock worth pennies, its CEO, a confidante of presidents, more or less evaporated, there must be lessons in there somewhere.

And indeed there is a lesson. Exclusive reliance on self-regulation is in and of itself insufficient to enforce higher corporate governance standards. For it has failed to deliver better governance in those markets where it has been applied longest. As Austin Mitchell, Member of Parliament for Great Grimsby, England, points out[29]:

There are too many instances of companies, rather like Vodafone [Enron, WorldCom, Tyco …] are run by a self-interested clique at the top, who owe a duty of care to no one, neither to stakeholders, employees or shareholders. That is why there are obscene salaries, with the gap widening between salaries at the top and those at the bottom. It is why we get such fat-cat pension schemes showering money on those who have got enough already. At the same time, there are raids on company pension schemes. Companies did that in the 1980s by taking pension holidays. In fact, they took £19 billion from company pension schemes by giving themselves pension contribution holidays. The moneys went directly into profits and into the pockets of the top board. There are also share options, and manipulation of moneys overseas—all being carried on at a time when companies are trying to increase their profits by downsizing, by transferring production overseas and by squeezing the salaries and conditions of their employees. It is an indecent spectacle.

What can we reasonably expect of corporate governance practice when companies are free to ignore codes of conduct at will? As Mitchell[30] makes clear, over-reliance on voluntary codes of conduct can lead to the all too familiar situation where "... scandals emerge only when a company goes belly up and everything emerges."

It is quite possible that gentleman's agreements between leading corporations about what constitutes good governance practices might have been well-suited to Victorian Britain when the club was small; but it was and remains a wholly inappropriate response to the competitive forces of modern international capitalism. It flies in the face of both the empirical evidence and rational reason that only when significant penalties exist for corporate malfeasance will sufficient attention be paid to corporate governance by board, senior management, and shareholders. Indeed, Sir Adrian Cadbury, one of the chief architects of self-regulation in the United Kingdom, searching for explanations for its failure, lamented[31]:

> The efficacy of the [gentleman's] club rules was routed in the self-interest of the membership in maintaining the reputation of the City and of their own firms within it. ... Those links were broken by a series of momentous changes. One was the sudden expansion of London's financial services sector in the 1980s. ... Old boundaries between different types of financial activity, with their differing rules, were swept away. ... Many new entrants to the City did not share the values of what they saw as the past. ... The gap in the framework of rules, which arose in the much enlarged City, was that nothing was put in place of the personal links with the heads of firms. There was no consistent means of passing on business values to newcomers and ensuring that they were adhered to.

The Role of Criminal Penalty

In recent years deceptive and fraudulent activities carried out by corporate officers and their senior employees have cost billions. Yet, Mitchell's *indecent spectacle* will continue especially where the consequences of executive malfeasance remain relatively minor. Take, for example, South Korea. During 1999, Daewoo, a global manufacturing company, was forced into bankruptcy with debts in excess of U.S. $70 billion. Kim Woochoong, the chairman, jumped ship—fleeing the country[32] and leaving his bewildered lieutenants to face massive public demonstrations.

Accusations of falsified company records used to conceal billions of dollars in losses were investigated by the South Korean authorities. By the end of 2001, 20 of Woochoong's executives had been successfully prosecuted and convicted. However, 13 of those convicted received suspended sentences.

This is not unusual—the British Broadcasting Corporation[33] reported in June 2004 that:

> The chairman of one of South Korea's largest conglomerates, SK Group, has been sent to prison for three years after embezzling company funds. Son Kil Seung, 63, was charged with illegally drawing more than $680m (£373m) from the group for private investment overseas. He was also found guilty of evading taxes and bribing politicians. In addition, he gave $8.7m in illicit campaign funds to the main opposition party ahead of the 2002 election.

Son Kil Seung's 3-year sentence was later suspended.[34]

It was the fictional character Gordon Gekko, a ruthless and greedy corporate raider played by Michael Douglas in the 1987 Oliver Stone film, *Wall Street*, who famously intoned "Greed is good." And in the United States, the rapacious greed exhibited by executive officers who plundered their companies' resources as freely as the Mighty Sparrow's celebrated drunk consumed rum, caused uproar in the general public and political establishment. The seemingly continuous "parade of catastrophe and disgrace" reported by the media during the first decade of the 21st century eventually forced the then president, George W. Bush, in his rather loquacious way to concede the need for change[35]:

> … we must usher in a new era of integrity in Corporate America. … My administration will do everything in our power to end the days of cooking the books, shading the truth and breaking our laws.

At the time, the creditability of President Bush's statement, in particular the commitment to real reform, was received with considerable skepticism. As Britain's *The Guardian* newspaper pointedly noted[36]:

> There is no mystery behind public reluctance to take the Bush team at its word. It is cut from the same shiny cloth as the corporate culture it now claims to abhor.
>
> A vivid illustration of the organic link between the administration and the companies now in the dock emerged this week. It appears that, in its efforts to push the election 2000 recount in Florida in Bush's direction, his campaign workers rushed about the state in corporate jets provided by three companies now under federal investigation—Enron, Halliburton and Reliant Energy. Not only do such uncomfortable facts present a serious credibility problem, it also means that no one in the administration dares take the microphone to provide leadership on economic issues.

Nevertheless, President Bush's statement did add considerable leverage to the voices calling for executives to be made more accountable for the criminal activities of their companies. The belief is the threat of criminal punishment will engender better governance and that[37]:

> If shareholders bear no responsibility for a manager's crime, they will have every incentive to hire managers willing to commit crimes on the corporation's behalf.

That corporate wrongdoing be adequately punished is a prerequisite for better corporate governance. While the nature of the punishment will vary from jurisdiction to jurisdiction,[38] there are important lessons to be learned from the American experience about both the timely execution of justice and severity of sentence imposed[39]:

> Effective regulators such as the SEC [U.S. Securities & Exchange Commission] can strike and strike hard. The Enron cases came to court within about five months of the company's collapse. The two men who were mostly involved have now gone on to serve long prison sentences. The same happened in the cases of WorldCom, Tyco, and other offenders. ... People should have the personal responsibility to pledge their position on the authenticity of the accounts of their company. That should be the responsibility of chief executives and chief financial officers.

DID THE CRIME? THEN SERVE THE TIME!

The air was thick with tension, courtroom packed, apprehension at what was about to happen:

> "Guilty." Judge Simeon T. Lake III's reading of the verdict landed like a bombshell in his federal courtroom in Houston. The first cries came from the second row, where the children of Kenneth L. Lay, the former Enron chairman, lurched forward and began sobbing.
>
> Dressed in a cobalt blue jacket, Linda Lay, Lay's wife, dropped her head onto his shoulder as the judge continued to read a series of fraud and conspiracy verdicts. Each count was punctuated by one word: "Guilty."
>
> When the judge finished, Lay, 64, had been convicted of 10 crimes—and a man who was once a close ally of President Bush and presided over one of the nation's most influential companies became someone who may spend the rest of his life in prison. (*The New York Times*, Sunday, June 4, 2006)[40]

The Benefit of Wolf Pack Capitalism

Mathematicians often talk about beautiful equations. Euler's identity where $e^{i\pi} + 1 = 0$ has a symmetry, elegance, and essence of the most profound. And the beautiful but infinitely complex unknown equation, which underlies capitalism, is just as profound. For it releases value to investors by virtue of the invisible hand of the market mechanism. It informs us that for good governance practices to take hold, investors and market participants must play an active role in scrutinizing the activities of corporate boards and senior management. Considerable value can be created when shareholders take on an activist stance.

Shareholder activism can be traced back to the late 1960s when socially responsible and environmentally sensitive movements began to shift the focus away from shareholder value toward measures of success, which included social and environmental considerations. For much of the late 1960s and 1970s, activist investors were at the fringes of capitalism, quietly ignored, as their investments represented a tiny fraction of outstanding corporate equity. Beginning in the early 1980s, pension funds for United States public employees and labor unions grew rapidly, providing a firmer foundation for an activist presence. By the end of the first quarter of 2010, Americans held around $16.5 trillion in retirement assets.

With the growth in assets under management, large institutional investors have increasingly played a more active role in monitoring and enforcing good governance standards in the corporations in which they invest. As owner-shareholders, they have the incentive and more importantly the ability to exercise closer oversight and control of management and corporate decision making. The trend is typified by Connecticut Treasurer Denise Nappier, the first black woman to be elected to serve as a state treasurer in the United States. Nappier, the principal fiduciary of the Connecticut Retirement Plans and Trust Funds, has used proxy votes to influence corporate governance practices on issues such as financial transparency and diversity within boards[41]:

> Think about it. If you had $20 billion to invest, you could do more than just make a whole lot of money. You could also create a wealth of positive change in the companies whose stocks you hold. At least that's how Connecticut State Treasurer Denise Nappier has been using her power. Entrusted with the pension funds of the state's 160,000 working and retired state employees, Nappier demands not only a good return, but good citizenship. Take American Electric Power, for example. For three years, Nappier sponsored shareholder resolutions urging the nation's largest producer of carbon dioxide emissions to study and report on the climatic impact of its electricity-producing plants. The company resisted, but Nappier persisted, and it finally agreed to conduct an analysis in 2005. Then there's Western food giant Safeway Inc.,

which Nappier challenged to appoint more independent board members when it found itself embroiled in an ugly labor dispute in 2003. ... The board has now added independent members. ... With power plays like these, Nappier, an African American native of Hartford, is helping shape policy inside the nation's traditionally clubby, white male corporate boardrooms.

The California Public Employees' Retirement System (CalPERS), which provides pensions to around 1.5 million retired state, school, and public sector employees in the state of California, is another long-standing active investor. In 1996, the CalPERS board developed a four-pronged approach to encouraging good governance in parts of its international portfolio:

1. Developing good governance principles.
2. Actively participating in local debates.
3. Proactive outreach to companies in their portfolio.
4. Strategy development with potential allies.

This strategy was actively pursued with differing degrees of success in the United Kingdom, France, Germany, and Japan.[42]

In recent years, hedge funds have become important players in the corporate governance debate. As returns in traditional asset classes have stagnated, capital has flowed into the hedge fund space. By the end of 2007, hedge funds managed in excess of U.S. $1.5 trillion. Hedge fund investors expect to generate large positive returns, which are uncorrelated with traditional asset classes. In return, they pay high management fees and share a proportion of their performance with the hedge fund. Typically, a hedge fund will charge a 2% management fee and a 20% performance fee. Given the fee structure, hedge funds are extremely aggressive in their search for performance. One of their strategies is to target a particular company by taking a large position in the stock. They then actively challenge managers to improve operating efficiency, business plans, and governance practices. When one hedge fund announces a significant position in a company, others will often quickly follow, forming a vicious *wolf pack* focused on maximizing shareholder value. As the influential financial periodical *Business Week* observed[43]:

Once they've got their teeth into a company, the new activists [hedge funds] usually won't let go. ... Such tenacity makes them formidable infighters ... [and they] are more dangerous to management than their predecessors. For starters, unlike mutual and pension fund managers, which often are trying to sell money management services to companies, the hedge funds are not tempted to pull their punches.

The agility of hedge funds and their ferocious promotion of shareholder value appear to have yielded some success[44]:

> Over the last few months, hedge funds have pressured McDonalds to
> spin-off major assets in an IPO; asked Time-Warner to change its busi-
> ness strategy; threatened or commenced proxy contests over H.J. Heinz,
> Massey Energy, KT&G, infoUSA, Sitel, and GenCorp; made a bid to
> acquire Houston Exploration; pushed for a merger between Euronext
> and Deutsche Boerse; pushed for changes in management and strategy
> at Nabi Biopharmaceuticals; opposed acquisitions by Novartis of the
> remaining 58% stake in Chiron, by Sears Holdings of the 46% minority
> interest in Sears Canada, by Micron of Lexar Media, and by a group of
> private equity firms of VNU; threatened litigation against Delphi; and
> pushed for litigation against Calpine that lead to the ouster of its top
> two executives.

The extent of boardroom capitulation in the face of *wolf pack capitalism* is
only now being documented, analyzed, and interpreted by academic scholars; for example, William Bratton, professor of the University of Pennsyvlania
Law School finds[45]:

> ... hedge funds have an enviable record success in getting targets to
> accede to their demands, using the proxy system with remarkable, per-
> haps unprecedented, success. If the pattern of intervention persists in
> time, expands its reach, and maintains the present high level of gover-
> nance success, then the separation of ownership and control becomes a
> less acute problem for corporate law.

The Inherent Ethos of Risk Management

Even today, if you say the phrase "corporate governance" to a senior executive, compliance usually comes to mind; risk management probably does
not cross their radar. It should have. Compliance alone is not enough; risk
management also has a critical role to play. The Economist Intelligence Unit
writing in the immediate aftermath of the Enron collapse recognized risk
management as one of the top 10 ways corporations could improve their corporate governance[46]:

> At the root of most company failures are ill-judged management deci-
> sions on risk. Non-executives need not be risk experts. But it is para-
> mount that they understand what the company's appetite for risk
> is—and accept, or reject, any radical shifts.

Regulators and legislators have long acknowledged this point. For example, Pat Barrett, while the auditor general for Australia, observed[47]:

Corporate Governance boils down to how an Organisation is managed, its corporate and other structures, its culture, its policies and strategies, and the ways in which it deals with its various stakeholders. ... The governance framework is concerned with structures and processes for decision making and with the controls and behaviours that support effective accountability for performance outcomes and results. This encompasses:

1. Defining and monitoring the strategic direction;
2. Defining policy and procedures to operate within the legal and social requirements;
3. Establishing control and accountability systems;
4. Reviewing and monitoring management and the organisation's performance; and
5 Risk management.

The key components of corporate governance in both the private and public sectors are business planning, internal controls including risk management, performance monitoring and accountability, and relationships with stakeholders. The framework requires clear identification and articulation of responsibility as well as a real understanding and appreciation of the various relationships between the organisation's stakeholders and those who are entrusted to manage resources and deliver required outputs and outcomes.

Successful companies are able to leverage their strengths, capitalize on competitor's weaknesses, exploit market opportunities, and run their operations efficiently and effectively. Risk management and robust internal controls are an essential element. Both are required for good corporate governance. As Alan Bollard, the then governor of the Reserve Bank of New Zealand, explained[48]:

In the financial system, corporate governance is one of the key factors that determine the health of the system and its ability to survive economic shocks. The health of the financial system much depends on the underlying soundness of its individual components and the connections between them—such as the banks, the non-bank financial institutions and the payment systems. In turn, their soundness largely depends on their capacity to identify, measure, monitor, and control their risks.

Responsible management will set up a system of corporate governance and controls that successfully identifies and manages the profit and potential losses associated with their business strategy. Clearly, risk management has an essential role to play. It can help prepare a company to manage potential threats and maximize opportunities to be gained from taking business risk. Although governance shortfalls can occur for many reasons, one of the more common shortcomings is focusing too much on the compliance of satisfying regulatory requirements without giving adequate attention to risk management and the internal control infrastructure and culture necessary to support and sustain a competitive edge.

Modern corporations must develop processes and operating structures that sustain good governance. Sustainability involves having robust risk management processes in place. The value of risk management in supporting this structure is most clearly demonstrated when one considers the core principles of risk management.

1. That someone has to be identified as being accountable.
2. That there exists in place a process to identify and evaluate risks.
3. There exist clear policies and procedures for managing risks.
4. There is a process in place to check that the policies and procedures are being adhered to.
5. If a business unit does not fully understand a risk, it must not engage in it, no matter what profits are claimed or reported—in the frequently recited adage: "Risk comes from not knowing what you're doing."[49]

Thus, we see risk management is much more than a technocratic compliance-type exercise; it also embodies significant values and ideals, not the least of which includes accountability and responsibility—the board of directors has an explicit responsibility to ensure all potential threats to a company have been systematically identified, carefully evaluated, and effectively controlled. The values implicit in risk management, as noted by London School of Economics Professor Michael Power, align closely with the principles underpinning good corporate governance[50]:

> Since the mid-1990s, risk management and private corporate governance agendas have become intertwined, if not identical. Since 1995 (the year of the collapse of Barings bank and of the Brent Spar crisis for Shell), being a "good" organisation has become synonymous with having a broad and formal risk management programme.

Business ethics are based on broad principles of integrity and fairness with focus on internal stakeholder issues such as product quality, customer satisfaction, employee wages and benefits, and so on. Risk management complements these core values[52]:

> From a management perspective, the recent incidence of high-profile scandals (Enron, Global Crossing, Tyco, inter alia) has produced a sharp portrait of the practical consequences of unethical conduct. Indeed, these consequences have led to impacts with which risk managers are quite

PRINCIPLES OF CORPORATE GOVERNANCE

The Organisation for Economic Co-operation and Development (OECD) has developed a set of principles for good corporate governance.[51] The guidelines consist of six principles:

Principle 1: Ensuring the Basis for an Effective Corporate Governance Framework:
> The corporate governance framework should promote transparent and efficient markets, be consistent with the rule of law, and clearly articulate the division of responsibilities among different supervisory, regulatory, and enforcement authorities.

Principle 2: The Rights of Shareholders and Key Ownership Functions:
> The corporate governance framework should protect and facilitate the exercise of shareholders' rights.

Principle 3: The Equitable Treatment of Shareholders:
> The corporate governance framework should ensure the equitable treatment of all shareholders, including minority and foreign shareholders. All shareholders should have the opportunity to obtain effective redress for violation of their rights.

Principle 4: The Role of Stakeholders in Corporate Governance:
> The corporate governance framework should recognise the rights of stakeholders established by law or through mutual agreements and encourage active co-operation between corporations and stakeholders in creating wealth, jobs, and the sustainability of financially sound enterprises.

Principle 5: Disclosure and Transparency:
> The corporate governance framework should ensure that timely and accurate disclosure is made on all material matters regarding the corporation, including the financial situation, performance, ownership, and governance of the company.

Principle 6: The Responsibilities of the Board:
> The corporate governance framework should ensure the strategic guidance of the company, the effective monitoring of management by the board, and the board's accountability to the company and the shareholders.

RISK MANAGEMENT IS NO SILVER BULLET

A robust risk management framework is no silver bullet. Well designed and rigorously enforced, it will provide a comprehensive approach to enhance overall corporate governance because:

1. Setting up a risk management framework provides a corporate-wide backdrop against which the corporate culture, business processes, and operational structures can be realigned toward effective management of potential opportunities while taking into account any associated adverse effects.
2. At the organizational level, it will help senior decision makers to think more strategically about the risks inherent in their business environment, thereby strengthening their ability to anticipate, assess, and manage risk.
3. It provides an organization with a mechanism to develop an overall approach to manage strategic and tactical risks by creating the means to discuss, compare, and evaluate risks from different business lines or activities on the same page.
4. It strengthens accountability and transparency and eases shareholder concerns by demonstrating that levels of risk associated with value-added activities are explicitly understood.
5. It enhances stewardship by strengthening the capacity of a business to safeguard assets and therefore long-term corporate survivability.

familiar: loss of reputation, legal penalties, dislocation of workers, drops in market value and credit rating, new legislation aimed at reforming the business world. Set in these terms, ethical considerations are very much like those conventional risks where human behavior and choice can lead to positive or negative outcomes. Ethics is a subject of both risk and risk management.

It is evident that risk management simultaneously respects and builds upon modern corporate values and ethical business practices while instilling respect for laws, internal regulations, and delegated authorities. In addition, it contributes to improved corporate citizenship, strengthened trust and transparency through sharing the results of risk analysis, and subsequent value-added actions amongst key decision-making employees. It is one element of the process by which companies can demonstrate commitment to the responsible maximization of shareholder value in line with increasing expectations of due diligence following intense media, shareholder, and regulatory scrutiny. It should be seen as an integral part of good management practice and sound corporate governance.

The Cost of Corporate Governance

While it is true that academics, regulators, and thoughtful corporations have long since acknowledged the value of good governance, it was only when Enron, WorldCom, and other failures put corporate governance on the front pages of our main newspapers that the wider business and political community began to focus serious attention on this issue. The interest has since been further spurred on by other well-publicized corporate abuses, mammoth investor losses, prison sentences for senior executives, and increasing vigilance by the regulatory agencies.

In the United States, the federal government has taken the lead role with legislation aimed at assisting the early identification of governance problems. For example, Section 112 of the Federal Deposit Insurance Corporation Improvement Act (FDICIA) of 1991 required institutions to have annual independent audits, assessments of the effectiveness of internal control over financial reporting, and independent audit committees. The Securities & Exchange Commission has also ensured that publicly traded companies provide detailed information about their operations, even before they are traded. Recent legislation aimed at improving the governance situation has focused on improving corporate disclosure.[53,54,55] The underlying rationale is that improvements in disclosures to shareholders and stakeholders produce credible publicly available firm-specific information and thereby serve to reduce information asymmetries, allowing investors to make more informed decisions about how to allocate their capital.

In the business community, widespread distrust of prescriptive legislation persists[56]:

> ASX's non-prescriptive approach to corporate governance is part of an overall focus on improving market efficiency and keeping agency costs low. The high standard of corporate governance practice disclosure in Australia has been achieved without the agency costs of "black letter" law common in other markets. Australia's regulatory package, including the cost of compliance, is directly linked to our attractiveness to global capital.

Indeed, the long-standing hostility to prescriptive legislation within the wider business community arose partly from the sense that efforts devoted to upgrading corporate governance are bureaucratic and costly, and partly from the erroneous but widespread notion that they add little value to the economy or firm.[57]

Corporate concern over the cost of implementing legislation aimed at improving transparency is not new. Yet, poor corporate governance at Enron alone took with it more than $60 billion in market value, almost $2.1 billion in pension plans, and 5,600 jobs. Settlement in the resultant fraud suit brought by the Securities & Exchange Commission in 2003 cost JP Morgan Chase $135 million, Citigroup $120 million, Merrill Lynch $80 million, and Canadian Imperial Bank of Commerce (CIBC) $80 million. Yet, these numbers pale into insignificance when one considers the settlement of class action lawsuits from defrauded Enron investors—Citigroup $2 billion, JP Morgan Chase $2.2 billion, and in August 2005 CIBC paid out $2.4 billion. The CIBC payout was around 10 times more than it had originally set aside for Enron litigation and one and one-third times its entire earnings in 2004.[58]

From a macroeconomic perspective, one of the factors, which is widely regarded as being essential to promote a healthy environment for long-term investment is good corporate governance.[59] An economic and regulatory system that promotes good corporate governance contributes to the attractiveness of a country in terms of inward investment and business development. It facilitates the operation of capital markets and the efficient flow of resources to investment opportunities. It is thus a critical ingredient in maintaining a sound financial system and a robust economy. As the Organisation for Economic Co-operation and Development makes clear[60]:

> How well companies are run affects market confidence as well as company performance. Good corporate governance is therefore essential for companies that want access to capital and for countries that want to stimulate private sector investment. If companies are well run, they will prosper. This in turn will enable them to attract investors whose support can help to finance faster growth. Poor corporate governance on the other hand weakens a company's potential and at worst can pave the way for financial difficulties and even fraud.

And this is why policy makers and regulators should take a close interest in corporate governance issues. Investment in good corporate governance is an essential investment not least because calls for more prescriptive legislation will continue to grow louder if policy makers believe there is widespread underinvestment by companies in their corporate governance infrastructure.[61]

From the perspective of the firm, it should be self-evident that sound corporate governance is essential to the long-term well-being of the company and its stakeholders, particularly its shareholders and creditors. Indeed, there is a growing body of empirical evidence, which suggests in countries where the policing of corporate governance issues is weak, the cost of capital is higher to reflect the additional risk.[62] Symbolic of this relationship is the situation that unfolded in Greece at the start of the second decade of the 21st century. Years of poor national governance manifested in unrestrained government spending left Greece with a 12.7% budget deficit in 2009. Prime Minister George Papandreou's socialist party found itself in a no-win situation of having to fend off bankruptcy by cutting the national deficit while attempting to keep electoral promises to help the poor. Rating agencies cut

PERFORMANCE AND GOOD GOVERNANCE

Of course, it needs to be stressed that good governance is not necessarily synonymous with good performance. In 1994, General Motors Corporation adopted a set of board governance guidelines. By 2004, its governance guidelines had evolved to include a 21-page set of ethical guidelines ranging from how to report wrongdoing to conflict of interest rules and insider trading issues. In October 2004, with its share price standing at around $38.5, it was the winner of Treasury and Risk Management's very first annual Corporate Governance Award for its long-established track record in innovative and effective governance practice.[63] Little more than one year later, by the middle of November 2005, its share price had fallen by over 41%.

the rating on Greek debt in December 2009 and then again to junk status in April 2010. With an increased cost of borrowing and unsustainable levels of debt, Papandreou was faced with the humiliation of going cap in hand to the International Monetary Fund. Thus, for both the wider economy and the firm, good corporate governance makes sound business sense.

Why Governance Failures Are Inevitable

Over a decade ago, governments around the world began to implement heavyweight legislation aimed at reforming corporate governance practices. In the United States, for example, the Sarbanes-Oxley Act was enacted in 2002 and contained significant penalties for internal control failures, including holding company directors personally responsible for accounting malfeasance. As a result, senior management in the United States, United Kingdom, and Europe are more aware of good governance practices than they were just a few years ago. For example, Lloyd's of London and the Economist Intelligence Unit found in the early part of the first decade of the 21st century that the time boards in global businesses spent on risk management rose fourfold.

It might be assumed that a company which introduces more formal audits of management performance, separates the positions of chairman and chief executive, appoints outside directors, and makes board members more accountable to shareholders has discharged its corporate governance responsibilities. It has not. This is made clear by Susan Bies[18]:

> On the one hand, we could pat ourselves on the back and say that things are generally going very well for most of the industry and we can finally

> tone down all of the corporate governance rhetoric. Or, we could say that
> those negative statistics apply only to the boards and senior managers
> at a small group of poorly rated institutions, which now have to pay the
> price. Or, yet again, we could say that effective corporate governance
> is a continuous process that requires ongoing vigilance on the part of
> the board, audit committee, senior management, and others within your
> bank. I hope you are thinking along the lines of this last sentiment.

Despite improvements in the regulatory framework, the constant stream
of corporate governance failures should continue to remind us that securing the consistent implementation of both the letter and spirit of good
governance is somewhat difficult to achieve. For sure, it depends on the
effectiveness of market regulation, the level of regulatory policing, and the
nature of enforcement. It also relies on the board's desire and skill in representing shareholder interests, which, in turn, rests partly on the board's
ability to monitor senior management. This ability can be easily compro-
mised as Jeffrey Gordon, Alfred W. Bressler Professor of Law at Columbia
University, observes[64]:

> Its board was a splendid board on paper, 14 members, only two insid-
> ers. Most of the outsiders had relevant business experience, a diverse
> set including accounting backgrounds, prior senior management and
> board positions, and senior regulatory posts. Most of the directors
> owned stock, some in significant amounts, almost all had received
> stock options or phantom stock as part of the director compensation
> package. ... The audit committee had a state-of-the art charter, attached
> to the 2001 Proxy Statement for all to admire, which made it the "over-
> seer of the Company's reporting process and internal controls" and
> gave it "direct access to financial, legal, and other staff and consultants
> of the Company" and the power to retain other accountants, lawyers,
> or consultants as it thought advisable. But if the report of the Enron
> Special Investigation Committee is accurate, the board was ineffec-
> tual in the most fundamental way, the Audit Committee particularly
> somnolent if not supine. It turns out that the independence of virtu-
> ally every board member, including audit committee members, was
> compromised by side payments of one kind or another. Independence
> was also compromised by the bonds of long service and familiarity. ...
> Things at Enron appeared to be going so well and management told
> such a convincing story that tell-tale signs of trouble—the proposal
> to suspend the corporate ethics code to permit conflicted transactions
> by a senior executive, an extraordinary request, really—didn't stir the
> antennae. Skepticism, suspicion, healthy scrutiny were inconsistent
> with the board's culture.

At Enron, rapacious greed fueled looting, bribery, and deception on an
unprecedented scale. However, even for well-run boards, danger continually
lurks, as Susan Bies also points out[18]:

As you know, once an organization gets lax in its approach to corporate governance, problems tend to follow. Many of you can recall the time and attention management devoted to Section 112 of FDICIA, which first required management reports and auditor attestations in the early 1990s. Then the process became routine, delegated to lower levels of management, and no longer relevant to the way businesses were being run. That is when the breakdown in internal controls began to occur. Unfortunately, trying to change the culture again is taking an exceptional amount of senior management and directors' time—time taken away from building the business. It is also taking more time from line managers and their staff. The challenge, therefore, is to ensure that banks' corporate governance practices keep pace with the changing risks that you will face in the coming years.

A common misconception is that corporate governance failures can be avoided by policy makers creating appropriate rules and regulators enforcing them. While legislation and regulatory mechanisms that seek to enforce good corporate governance are necessary, they are simply not sufficient to establish and entrench corporate accountability and responsibility. This is because, as George Orwell captures in these two passages from his shadowy poem "The Lesser Evil," despite being "dark & mean," the "house of sin," for many individuals, has an irresistible allure:

> The house of sin was dark & mean,
> With dying flowers round the door;
> They spat their betel juice between
> The rotten bamboos of the floor.
> Why did I come, the woman cried,
> so seldom to her beds of ease?
> When I was not, her spirit died,
> And would I give her ten rupees.

Future breakdowns in corporate governance resulting in financial distress to shareholders, stakeholders, and even the wider economy are inevitable because rampant greed and cupidity are inherent human characteristics. Adam Smith, in what many consider to be his greatest work, observed[65]:

> There are two different occasions upon which we examine our own conduct, and endeavour to view it in the light in which the impartial spectator would view it: first, when we are about to act; and secondly, after we have acted. Our views are apt to be very partial in both cases; but they are apt to be most partial when it is of most importance that they should be otherwise. … So partial are the views of mankind with regard to the propriety of their own conduct, both at the time of action and after it; and so difficult is it for them to view it in the light in which any indifferent spectator would consider it. But if it was by a peculiar faculty, such as the

moral sense is supposed to be, that they judged of their own conduct, if they were endued with a particular power of perception, which distinguished the beauty or deformity of passions and affections; as their own passions would be more immediately exposed to the view of this faculty, it would judge with more accuracy concerning them, than concerning those of other men, of which it had only a more distant prospect.

This self-deceit, this fatal weakness of mankind, is the source of half the disorders of human life. If we saw ourselves in the light in which others see us, or in which they would see us if they knew all, a reformation would generally be unavoidable. We could not otherwise endure the sight.

In individuals such as Kenneth Lay, Jeffrey Skilling, and Andrew Fastow,[xlii.] self-deceit is so complete that even when caught and convicted they are unable to acknowledge their error:

I have gone back and tried to think what I would have done differently given the facts at the time. And quite frankly, there is nothing I can come up with. (Jeff Skilling, president and chief operating officer of Enron)[66]

While one might hope the temptation to indulge in poor governance practices is tempered somewhat by the consequences of being caught, corporate leadership and responsibility brings with it extraordinary temptation. As William Bernstein observes, this circumstance not only attracts to the position scoundrels, but also turns people who should know better into scoundrels[67]:

When he began his career in finance, no one would have picked out Michael Smirlock as a future felon. Brought up in a household presided over by, in the words of one of his friends, a "classic Jewish intellectual" father, Michael excelled academically and acquired a Ph.D. in finance. Six years later, he was awarded a tenured chair at the Wharton School at the University of Pennsylvania. ... Drawn by the lure of bigger money, he found himself in 1990 at Goldman Sachs; by 1992, he had made partner. The very next year, he garnered a $50,000 fine and three month suspension ... for late-trade allocations and was forced to resign. He then set up a real estate investment trust and a series of hedge funds. On 24 May 2002, Judge Gerald E. Lynch of the Federal District Court for the Southern District of New York sentenced him to four years incarceration and fined him $12.6 million for fraudulently concealing losses from his investors ... if this highly respected academic, who should not have had any problem with the legal and ethical concepts involved, could not keep his hands out of the cookie jar, what chance does the average broker or B-school grad have?[68]

That even those individuals who should know better do not is perhaps a powerful testimony to life coach guru Napoleon Hill's assertion[69]:

Nearly all forms of lower animals have instinct but appear not to have the power to reason and think; therefore, they prey upon one another physically. Man, with his superior sense of intuition, thought and reason, does not eat his fellow men bodily; he gets more satisfaction out of eating them FINANCIALLY! ... for there is food and shelter and raiment and luxury of every nature sufficient for the needs of every person on earth, and all these blessings would be enjoyed by every person except for the swinish habit that man has of trying to push all the other "swine" out of the trough, even after he has all and more than he needs.

That unscrupulous men will prey financially upon their fellow man unless impeded is why ultimately, as David Nadler, vice chairman of the global professional services firm Marsh & McLennan Companies, makes clear, it is up to the board to address gross deficiencies in management practices, ethics, and corporate responsibility[70]:

The key to better corporate governance lies in the working relationships between boards and managers, in the social dynamics of board interaction, and in the competence, integrity and constructive involvement of individual directors.

Risk management will fail without strong governance. Problems with corporate governance start at the board of directors because it has final responsibility for the functioning of the firm. It is therefore critical the directors, senior executives, investors, and other interested parties remain vigilant in enforcing good governance and in that task understand the supporting role of risk management. Yet, we must acknowledge that there are some things you cannot legislate, risk manage, or teach in business school. Sound business judgment and integrity are two of those things. Had they been in greater supply, the corporate landscape, and the fate of companies like Lehman Brothers, IndyMac, and Enron, might have been quite different. The plain and simple fact of the matter is that unscrupulous managers and boards can still—and always will—use their influence and control over corporations to benefit themselves at the expense of their shareholders, creditors, employees, and other stakeholders. And this, as distasteful as it may seem, is a fundamental rule of risk management.

For Further Thought

This chapter has not discussed at any length specific corporate governance codes. This is because most countries with developed economies have published governance codes specific to their own markets. These codes tend to

KEY POINTS

Key Point 1: The key governance problem is the separation of ownership and control. Good corporate governance flows from a well-functioning system of corporate direction, regulatory oversight, and control through the market mechanism.

Key Point 2: Do not make the mistake of seeing corporate governance solely as a compliance exercise. Any institution that takes this view is missing the opportunity to create and sustain a competitive edge. At worst, they may result in value destroying behavior where compliance departments indulge in a paper exercise, ticking boxes rather than doing anything productive.

Key Point 3: If you say the phrase *corporate governance* to a senior executive, compliance usually comes to mind; risk management probably did not cross their radar. It should have because at the root of most company failures are ill-judged management decisions on risk.

Key Point 4: Risk management is much more than a technocratic compliance type exercise; it also embodies significant values and ideals, not least of which are accountability and responsibility. It builds upon corporate values and ethical business practices while instilling respect for laws, internal regulations, and delegated authorities. Effective risk management contributes to improved corporate citizenship, strengthened trust, and transparency.

Key Point 5: Strong corporate governance is synonymous with having a broad, well understood, and widely accepted risk management process.

Key Point 6: Securing the consistent implementation of both the letter and spirit of good governance principles depends on the effectiveness of market regulation, the level of regulatory policing, nature of enforcement, and explicit commitment of senior management.

Key Point 7: Future breakdowns in corporate governance resulting in financial distress to shareholders, stakeholders, and the wider economy are inevitable because greed and cupidity are inherent human characteristics. Therefore, effective corporate governance is a continuous process that requires a commitment to ongoing vigilance.

Key Point 8: An economic and regulatory system that promotes good corporate governance contributes to the attractiveness of a country, lowers the cost of capital, and facilitates the operation of efficient capital markets.

be enshrined in national laws, securities laws, and/or stock exchange listing standards.[71] The commonality between codes lies in their acknowledgement that effective corporate governance will comprise a system of internal controls, checks and balances, and independent external verification whereby inappropriate activities can be prevented, exposed, and punished.

Some basic issues boards, senior executives, management, scholars, and active investors might like to consider include:

- The adoption of formal principles of corporate governance.
- The number of independent directors.
- Whether outside members of the board meet in executive session without any members of management.
- The frequency of board meetings.

Increasingly, a basic framework of corporate governance is associated with the presence of all of the following:

1. Separate chairman and CEO.
2. Lead independent director.
3. Independent audit committee.
4. Independent compensation committee.
5. Independent corporate governance committee.
6. Corporate governance guidelines approved by the board.
7. Outside directors hold meetings without management present.
8. Annual board self-evaluation.
9. Annual review of independence of board.
10. Charters for audit, compensation, and corporate governance committees.
11. Charter for lead independent director.
12. Board orientation/education program.
13. Corporate compliance program.
14. Disclosure committee for financial reporting.
15. Code of ethics.

Specific questions for further discussion include:

1. Is the board of directors dominated by insiders?
2. Are board activities dominated by the presence of a *celebrity* CEO?
3. Does the company have its own best practice template (or follow specific best practice principles) that go above and beyond those mandated by regulatory bodies?
4. Is there an independent committee to nominate directors?

5. Is there a reasonable degree of director turnover? Are new voices encouraged to provide a fresh perspective on the board? Or is it full of *old-timers*?

6. What is the educational background and experience of board members?
   - Do they have adequate business experience?
   - Are they members due to political or other influence?

7. What is the depth of financial expertise of board members on the audit committee?

8. What is the level of director absenteeism?

9. What is the implementation plan of the ethics policy? Is the policy coherent and implementation clear?

Additional questions for further thought include:

1. Does your organization have clarity regarding roles and responsibilities for risk management and compliance?
   - Who is the champion of risk management?
   - Are they seen as separate elements of the internal control mechanism?
   - Who does each report up to?
   - What is the value proposition of each?

2. How is the efficiency and effectiveness of compliance measured?
   - Are the metrics appropriate?
   - Who is monitoring them?
   - What are the appropriate and acceptable performance parameters?
   - What are the consequences of failure to perform?

3. Who are the various stakeholders that should have an interest in the performance of compliance and risk management?
   - What is their level of seniority?
   - Do they carry enough weight in the organization to be taken seriously?
   - Are they actively engaged?

Additional Resources

The essence of the corporate governance problem is discussed in the classical work of Berle and Means (1932). Corporate governance codes of practice are discussed in CalPERS (1996), Cadbury (1998), the Committee on Corporate Governance (1998), Davies (2002) and the Australian Securities Exchange Corporate Governance Council (2007), and Sanford (2007). Davies (2002) and Bhattacharya and Daouk (2002) discuss the impact on the cost of equity of governance practices. The World Council for Corporate Governance (www. wcfcg.net) and Transparency International (www.transparency.org) are two leading international organizations engaged in promoting good governance practices worldwide. In addition, Institutional Shareholder Services publishes a broad range of position papers, articles, newsletters, academic studies and other research surrounding corporate governance issues and best practices (www.issproxy.com). The use of accounting devices to avoid disclosing sizeable losses is discussed in Melis (2005). The role of hedge funds in corporate governance is touched on in Kahan and Rock (2006) and Bratton (2007). Posner (1986) outlines shareholder incentives for good governance. Lang and Jagtiani (2010) discuss the role of risk management and corporate governance within the context of the global financial crisis. They focus in particular on the onset of the financial crisis in August 2007 with the collapse of the asset-backed commercial paper market. The role of the principal-agent problem and the breakdown in corporate governance is traced back to the collapse of large financial firms.

Key regulatory speeches outlining issues in corporate governance, compliance, and risk management are given by Barrett (2001), Bollard (2003), Bies (2004), Cox (2007), and Paredes (2010). Key findings related to compliance and risk management are discussed in Economist Intelligence Unit (2002), PricewaterhouseCoopers and the Economist Intelligence Unit (2004), and Lloyd's of London and Economist Intelligence Unit (2005). Some of the challenges facing compliance are outlined in Parker and Lehmann (2006). The central role of risk management is discussed in Gordon (2002), Nadler (2004), and Power (2004). Ethical issues are touched upon in Smith (1759), Young (2004), Bernstein (2006), and Oberlechner (2007).

Much has been written on specific corporate governance failures. Articles that have stood well over the passage of time include:

- Davies (1995) and Bower (1996) who discuss the case of Press Baron Sir Robert Maxwell.
- The demise of the investment firm Barlow Clowes is reviewed in Hamilton (1990), Johnson (1989), May and Vaughan (1988), Office of the Parliamentary Ombudsman (1989), and Younghusband (1990).
- The Guinness affair is discussed at length in Kochan and Pym (1987).
- O'Rourke (2005) outlines the infamous Daewoo saga.

Australian Securities Exchange Corporate Governance Council. (2007). *Corporate Governance Principles and Recommendations*, 2nd ed. Australia: ASX Corporate Governance Council.

Barrett, P. (2001). "Corporate Governance: More Than Good Management." Address at the CPA South Australia Annual Congress, Adelaide, Australia, November 16.

Berle, A. and Means, G. (1932). *The Modern Corporation and Private Property.* New York: Commerce Clearing House.

Bernstein, W.J. (2006). Corporate Finance and Original Sin. *Financial Analysts Journal* 62(3) (May/June):20–23.

Bhattacharya, U. and Daouk, H. (2002). The World Price of Insider Trading. *Journal of Finance* 57:75–108.

Bies, S.S. (2004). "Trends in Risk Management and Corporate Governance." Address at the Financial Managers Society Finance and Accounting Forum for Financial Institutions. Washington, DC, June 22.

Bollard, A. (2003). Address to the Annual Meeting of the Institute of Directors in New Zealand, Christchurch, April 7.

Bower, T. (1996). *Maxwell: The Final Verdict.* London: Harper Collins.

Bratton, W. (2007). Hedge Funds and Governance Targets. *European Corporate Governance Institute Working Paper Series in Law*, Working Paper 80/2007(February).

Cadbury, A. (1998). The Future for Governance: The Rules of the Game. *Journal of General Management* 24(1) Autumn:1–14.

CalPERS. (1996). CalPERS Adopts International Corporate Governance Program. March 18. CalPERS press release.

Committee on Corporate Governance. (1998). *Final Report.* Governance and. London: Gee Publishing Ltd.

Cox, C. (2007). Address to the 2007 US-EU Corporate Governance Conference, Washington, DC, October 9, 2007.

Davies, H. (2002). Address at the China Securities Regulatory Commission on Corporate Governance and the Development of Global Capital Markets Meeting, April 22, 2002. Full speech http://www.fsa.gov.uk/Pages/Library/Communication/Speeches/2002/sp96.shtml.

Davies, R. (1995). *Foreign Body: The Secret Life of Robert Maxwell.* London: Bloomsbury.

Economist Intelligence Unit. (2002). *Corporate Governance: The New Strategic Imperative.* London: The Economist Intelligence Unit.

Gordon, J.N. (2002). What Enron Means for the Management and Control of the Modern Business Corporation: Some Initial Reflections. *University of Chicago Law Review* 69.

Hamilton, S. (1990). Barlow Clowes: Who's to Blame. *The Accountant's Magazine* (March): 28–29.

Hill, N. (1928). *The Law of Success.* Meriden, CT: The Ralston University Press.

Jacoby, S.M. (2007), Principles and Agents: CalPERS and Corporate Governance in Japan. *Corporate Governance: An International Review* 15:5–15. doi: 10.1111/j.1467-8683.2007.00537.x.

Johnson, H. (1989). The Barlow Clowes Affair and Government Regulation. *International Banking Law* 7(8):114–115 .

Kahan, M. and Rock, E.B. (2006). Hedge Funds in Corporate Governance and Corporate Control. *New York University Law and Economics Working Papers*, Paper 68.

Kochan, N. and Pym H. (1987). *The Guinness Affair: Anatomy of a Scandal.* London: Christopher Helm Publishers Ltd.

Lang, W.W. and Jagtiani, J.A. (2010). The Mortgage and Financial Crises: The Role of Credit Risk Management and Corporate Governance. *Atlantic Economic Journal* 38(2):123–144.

Lloyd's of London and Economist Intelligence Unit. (2005). *Taking Risk on Board: How Global Business Leaders View Risk*. London: Lloyd's of London.

May, P. and Vaughan, S. (1988). The Barlow Clowes Affair. *International Banking Law* 7(3):34–38.

Melis, A. (2005). Corporate Governance Failures: To What Extent Is Parmalat a Particularly Italian Case? *Corporate Governance: An International Review* 13(4):478.

Nadler, D.A. (2004). Building Better Boards. *Harvard Business Review* 82:102–111.

Paredes, T. (2010). Remarks at the 22nd Annual Tulane Corporate Law Institute, Washington, DC, April 15.

Oberlechner, T. (2007). *The Psychology of Ethics in the Finance and Investment Industry*. Research Foundation of the CFA Institute, Charlottesville, VA. http://www.cfapubs.org.

Office of the Parliamentary Ombudsman. (1989). *First Report—Session 1989–1990: The Barlow Clowes Affair*. London: HMSO.

O'Rourke, M. (2005). Fugitive Daewoo Executive Returns. *Risk Management Magazine* (August). Risk and Insurance Management Society.

Parker, C. and Lehmann, N.V. (2006). Do Businesses Take Compliance Systems Seriously? An Empirical Study of the Implementation of Trade Practices Compliance Systems in Australia. *Melbourne University Law Review* 30.

Posner, R. (1986). *The Economic Analysis of Law*. Chicago: University of Chicago Press.

Power, M. (2004). *The Risk Management of Everything: Rethinking the Politics of Uncertainty*. London: Demos.

PricewaterhouseCoopers and the Economist, London Intelligence Unit. (2004). *Governance: From Compliance to Strategic Advantage*. The Financial Services Authority (www.fsa.gov.uk), Financial Reporting Council (www.frc.org.uk), and the Securities & Exchange Commission (www.sec.gov), offer additional information on the economic role and value of corporate governance.

Sanford, J.M. (2007). Principles and Agents: CalPERS and Corporate Governance in Japan. *Corporate Governance: An International Review* 15(1):5–15.

Smith, A. (1759). *The Theory of Moral Sentiments*. London.

Young, P.C. (2004). Ethics and Risk Management: Building a Framework. *International Journal of Risk Management* Spring.

Younghusband, V. (1990). Financial Regulation: The Barlow Clowes Affair. *Journal of International Banking Law* 5(3):76–79.

Endnotes

1. The Mighty Sparrow is a calypso singer whose real name is Slinger Francisco.
2. Soca, or soul of calypso, is an upbeat fast-paced derivative of calypso music. It is perhaps best characterized by Alphonsus Celestine Edmund Cassell's 1982 international hit, "Hot, Hot, Hot."
3. *Seattle Times*, December 29, 2002.

4. See, for example, CNNMoney (http://money.cnn.com/2009/10/15/real_estate/foreclosure_crisis_deepens/) or RealtyTrac (http://www.realtytrac.com/).

5. He was sentenced to 50 years in prison. The case is *USA v. Petters et al.*, U.S. District Court, District of Minnesota, No. 08-00364.

6. See U.S. Securities & Exchange Commission Litigation Release No. 21255/October 16, 2009. *SEC v. Galleon Management, LP, Raj Rajaratnam, Rajiv Goel, Anil Kumar, Danielle Chiesi, Mark Kurland, Robert Moffat, and New Castle LLC,* Civil Action No. 09-CV-8811.

7. See *United States v. Robert Allen Stanford et al.* Court Docket Number: H-09-342.

8. Directed at Bank of America and posted in a YouTube video which went viral. See further details at the *Huffington Post* (http://www.huffingtonpost.com).

9. On the 29th of April to be exact.

10. The case is *Securities & Exchange Commission v. Goldman Sachs & Co. and Fabrice Tourre*.

11. See Carl Levin, U.S. Senator, Michigan (http://levin.senate.gov/newsroom/release.cfm?id=324169).

12. This was the case even though Douglas Hogg was a member of the opposition Conservative party rather than the governing Labour party.

13. See Permanent Subcommittee on Investigations (http://hsgac.senate.gov/public/index.cfm?FuseAction=Subcommittees.Investigations).

14. In all of these cases, managers (of the Energy Firm, Enron, Telecommunications firm, WorldCom and the Italian food giant, Parmalat) allegedly used various accounting devices to avoid disclosing sizeable losses, possibly with the collusion of at least some auditors and lawyers. See Melis (2005).

15. Notice that in the United States, one might expect the market for corporate control, manifest in takeovers, to provide a powerful incentive toward good corporate governance. This is less so in the United Kingdom where the market for corporate control has been historically less aggressive.

16. See Berle and Means (1932).

17. See, for example, Bies (2004) and PricewaterhouseCoopers and the Economist, London Intelligence Unit (2004).

18. See Bies (2004).

19. See Parker and Nielsen (2006).

20. In 1988, the investment firm Barlow Clowes collapsed after it emerged that chairman Peter Clowes and his associates had misappropriated in excess of £100m from small, private, and elderly investors, who believed they were investing in government securities. Sixteen investors took their own lives in the aftermath of the fraud and Peter Clowes was imprisoned for his actions. See Hamilton (1990), Johnson (1989), May and Vaughan (1988), Office of the Parliamentary Ombudsman (1989), and Younghusband (1990).

21. In a compelling drama of human weakness—of greed, of cupidity, of misplaced trust, and of injurious human loss: the loss of jobs, of savings, of reputations and of lives; Press Baron Sir Robert Maxwell systematically plundered his companies' pension funds to finance complex corporate deals. Following his bizarre death in 1991, the financial problems were exposed—debts of around £4 billion and a £441 million hole in the pension funds. See Bower (1996) and Davies (1995).

22. A share price manipulation scandal to drive up the price of Guinness shares during a takeover battle for the Scotch whisky company distillers in 1986. See Kochan and Pym (1987).
23. See Davies (2002).
24. See the Committee on Corporate Governance (1998).
25. See Corporate Governance Principles Press Release (http://www.asx.com.au/documents/about/mr20070802_revised_corporate_governance_principles.pdf).
26. See Australian Securities Exchange Corporate Governance Council (August 2007).
27. Speaking at the U.S. Senate Commerce Committee hearing on the Enron bankruptcy in February 2002. For more details, see the *Houston Chronicle* (http://www.chron.com).
28. See Daniel Henninger, *The Wall Street Journal*, February 8, 2002.
29. Second reading of Company Law Reform Bill, *Commons Hansard Debates*, June 6, 2006.
30. Ibid.
31. See Cadbury (1998).
32. He returned to South Korea in 2005, apparently tired of life as a millionaire on the run. He was promptly arrested by the authorities.
33. See *BBC News* (http://news.bbc.co.uk/2/hi/business/3845627.stm).
34. See O'Rourke (2005).
35. President George W. Bush (July 9, 2002).
36. See *Guardian Unlimited*, Wednesday August 7, 2002.
37. See Posner (1986).
38. For example, Enron's Jeffrey Skilling was accused of breaching his fiduciary responsibility in exchange for salary, bonuses, and other compensation he received through a scheme that artificially inflated Enron's share price.
39. Austin Mitchell, *Commons Hansard Debates*, June 6, 2006.
40. In one final twist of fate, Lay escaped life in prison. For on July 6, 2006, he suffered a massive heart attack and died at his holiday home near Aspen, Colorado. Reflecting the widespread anger that right up until the end he never acknowledged his culpability, the *New York Post* screamed: *"LAY HIM LOW: ENRON'S CHIEF CROOK DUCKS BIG HOUSE BY DROPPING DEAD."* Following a few days of reflection, Joe Nocera writing in the *New York Times* under the title "Even up to His Dying Day, Enron's Lay Didn't Get It," neatly summed up the legacy of the convicted felon:" ... Yes, there were others more directly responsible for Enron's collapse, like its former chief financial officer, Andrew S. Fastow, whose devious partnerships both allowed Enron to disguise the truth of its financial condition and then sowed the seeds of its destruction. The former Enron president Jeffrey K. Skilling, too, played a greater role in Enron's collapse, since he was the company's hands-on leader, and set the tone that made greedy behavior and shady accounting standard operating procedure ... Thanks in large part to Enron and Lay, much has changed in corporate America. Boards are under far more scrutiny than they used to be. We now have Sarbanes-Oxley, a law that has stiffened the spine of the nation's accountants, and forced companies to spend millions on fraud prevention measures. Chief executives now have to sign their companies' financial statements, asserting to their accuracy. Many business schools have placed a new emphasis on ethics." *New York Times*, July 9, 2006.

41. *Ms. Magazine*, Spring 2005.
42. See CalPERS (1996) and Jacoby (2007).
43. *Business Week*, February 20, 2006.
44. See Kahan and Rock (2006).
45. See Bratton (2007).
46. See the Economist Intelligence Unit (2002).
47. See Barrett (2001).
48. See Bollard (2003).
49. This saying is frequently attributed to Warren Buffett (most noted for his Berkshire Hathaway investment company).
50. See Power (2004).
51. Endorsed by OECD Ministers in 1999 and revised in 2004, they have become the leading international corporate governance benchmark for policy makers, investors, and corporations. For further details see Organisation for Economic Co-operation and Development (www.oecd.org).
52. See Young (2004).
53. Sarbanes-Oxley Act in the United States, the Law on Financial Security in France, and Kontrag in Germany all require public companies to be more transparent about their management of business risk. In the United States, for example, the Sarbanes-Oxley Act was enacted in 2002. Section 302 of the Act requires executive representations by certifying officers and Section 404 requires an annual assessment of the effectiveness of internal control over financial reporting. It requires documenting financial reporting policies and procedures and the relevant controls; and evaluating the controls design and controls operating effectiveness. The Act contains significant penalties for internal control failures including holding company directors personally responsible for accounting malfeasance.
54. Such as the use of accounting devices for the sole purpose of obscuring adverse results.
55. Notice that disclosure is not a new approach. In 1993, the Federal Deposit Insurance Corporation (FDIC) in the United States issued regulations under Section 36 requiring every insured depository institution with $500 million or more in total assets to submit to the FDIC and other appropriate federal and state supervisory agencies an annual report that included audited financial statements, a statement of management's responsibilities, assessment by management of the effectiveness of internal control over financial reporting and compliance with designated laws and regulations, and an auditor's attestation report on internal control over financial reporting, and that each regulated institution establish an independent audit committee of its board of directors comprised of independent outside directors.
56. Eric Mayne, Chief Supervision Officer, ASX, 2 AUGUST 2007. See ASX (www.asx.com).
57. See, for example, Davies (2002).
58. In 2004, CIBC reported $1.8 billion in earnings.
59. See, for example, the annual Global Competitiveness Report produced by World Economic Forum (www.weforum.org). The survey pulls together the key factors for national competitiveness.
60. See Organisation for Economic Co-operation and Development (http://www.oecd.org).

61. Or future failures have a significant impact on investors or the wider economy.
62. For example, Davies (2002) and Bhattacharya and Daouk (2002) among others found the cost of equity for firms is between 3% to 5% higher than in countries where insider dealing is policed effectively.
63. See Treasury and Risk (www.treasuryandrisk.com).
64. See Gordon (2002).
65. See Smith (1759).
66. Speaking at the U.S. Senate Commerce Committee hearing on the Enron bankruptcy in February 2002. For more details see the *Houston Chronicle* (http://www.chron.com/disp/story.mpl/special/enron/1272993.html).
67. See Bernstein (2006).
68. Smirlock spent 37 months in a federal prison camp after pleading guilty to securities fraud. On his release and permanently changed for the better by his experiences while imprisoned, he became Executive Director of STRIVE, an organization devoted to securing jobs for the chronically unemployed—see STRIVE (http://www.strivenational.org/).
69. See Hill (1928).
70. See Nadler (2004).
71. Details of governance codes for much of the developed and developing world: see the European Corporate Governance Institute Web site (www.ecgi.org).

5

The Most Important Lesson a Risk Manager Must Know

I was approached one day at a conference by an unknown young man who had recently attended one of my talks. He was giddy with excitement, it being his first investment conference of this type. He had had the opportunity to approach and ask questions of some of the greatest minds in the investment world, men and women who are titans of our time. Now, I guessed, he was turning his attention to the minnows, namely me. Yet, he brimmed with enthusiasm, fascinated to hear what I had to say, to absorb every word. He began with the phrase, "I am new to risk management, and I know you are extremely busy, but can I take a moment of your time to ask you one question?"

I was indeed in quite a hurry (to catch a train). I paused, indicating my acceptance. "One question only, then I must dash," I stated as politely as I could given the time pressure.

"Thank you," said the young man, "I really enjoyed your talk. Can you tell me what you consider to be the most important lesson for risk managers?"

I scratched my nose, smiled, then replied with one of my favorite stories.

The story involved a man by the name of Richard Munslow. Munslow was a sin eater. In actual fact, he was the very last sin eater. And he lived on the Marches, that slip of land, dotted with mountains, moorlands, green wooded valleys and castles, which separates England from the Principality of Wales. Sin eaters were always few in number and one imagines this had to be due to the nature of their work. Detested by the local population, they were an essential element of the funeral custom of the region. The sin eater would be summoned on the death of a loved one who had either refused or was unable, due to the sudden nature of their demise, to recant their sins. His function was to dine on beer or wine, and bread often served with cheese or salted meat. This might sound quite appealing to you. Indeed, it might have been, if it were not for the distasteful fact the meal had to be consumed directly off the decaying corpse of the unrepentant sinner. This, it was believed, would transfer the sins from the soul of the recently departed to the soul of the consumer of the beer, bread, and cheese—the sin eater.[1]

Unthinking risk managers can find themselves in a similar position to the detested sin eater. Tempted by the prospect of *beer, bread, and salted meat,* many unconsciously straddle the polarity between passionate, risk-seeking rain makers and the masked despair, calamity, and legal penalty that result from ignoring risk or breeching risk appetite. Take for example, Robert Jones, the chief risk officer of Amaranth Advisors LLC. He was reported to have been paid a bonus of around U.S. $5 million for 2005.[2] In September 2006, the hedge fund lost around U.S. $6 billion on natural gas investments. At the time, it was the largest hedge fund collapse in history. Just before the calamity, on August 29th, the chief executive officer, Nick Maounis, commenting on the firm's investment strategy stated:

> Spreads and options are of their very nature instruments for positions which are designed to allow the user to capture upside with a much clearer understanding with respect to downside exposure.[3]

In other words, the strategy was supposed to minimize risk and maximize reward. Yet, Amaranth's risk management system, headed by Jones, did not appear to measure correctly how much risk his firm was facing. It was as if the chief risk officer was missing in action precisely at the moment his presence was most required. Jones, stomach bloated from an ample supply of compensatory *beer and bread* failed to step *up to the risk management plate*. Where was the risk mitigation plan and requisite action that would limit losses arising from the firm's highly concentrated natural gas bets? It is doubtful Jones fully comprehended the significance of such concentrated positions. For if he had, and had he acted effectively on his concerns, the outcome could have been very different. As Marc Freed, a managing director at a hedge fund advisory firm, stated in total astonishment[3]:

> It was a total failure of risk control to put your entire business at risk and not seem to know it.

That risk managers should be sought out by executive leaders solely to absorb the *sins* of a risk-ignoring venture is lamentable. That well-educated, intelligent individuals should willingly conform to the role disappoints.[4] At the energy company Enron, Rick Buy headed up the risk assessment and control (RAC) department. Appointed in 1999, he had a specific mandate to monitor deals to protect Enron's interests. His RAC department purported to conduct sophisticated, minute-by-minute mathematical analyses to assess the risk in Enron's various deals.[5] But his teams' muttered incantations about excessive risk taking were largely ignored by senior management. It appears Buy was able to overlook this slight and focus instead on the *beer, bread, and salted meat* rather than the stench emanating from the rotting corpse that was Enron[6]:

> HOUSTON, March 1—A former Enron managing director who evaluated risk for the company testified on Tuesday that he strenuously objected to the formation of off-balance-sheet partnerships [These partnerships allowed Enron to conduct transactions off of its books and also enabled Enron to avoid reporting losses.] directed by Andrew S. Fastow and urged the company to "come clean" about the risks they posed. ... Mr. Kaminski is one of the first key witnesses for the government

who is not testifying under a government cooperation agreement. The Polish-born Mr. Kaminski, 58, worked for Enron for a decade and now heads the quantitative risk management team at Citigroup's energy trading group. He said that in early June 2001, he was approached by Mr. Skilling and others about wanting to use LJM [an off-balance-sheet partnership] to hedge, or insure against loss, an investment in Rhythms Net Connections, an Internet start-up. The transaction involved donating Enron shares to LJM and using the shares to back the investment. After studying the idea, Mr. Kaminski said, he told Richard Buy, Enron's chief risk officer, that he opposed the deal, comparing it to "gambling in a casino that is insolvent." As Mr. Kaminski saw it, the entire structure depended on Enron's stock going up or staying even—a risk he considered too great, he said. The deal was nevertheless approved and Mr. Kaminski got the call a month later from Mr. Skilling, who pushed his team out of the risk department and into the wholesale energy division.

The most disturbing thing about all of this is the level of contempt for risk exhibited by those with a clear mandate to oversee. Yet, contempt for risk has been an enduring theme in almost every area of the financial services industry. Should one be surprised that it is also exhibited by the overseer of risk? Take for example, the calamitous decline of the once great firm IndyMac. Headquartered in Pasadena, California, IndyMac Bank Corporation was once one of the largest originators of mortgage loans in the United States. Its chairman and chief executive officer was a man by the name of Michael Perry. His business model appeared to be largely built upon the erroneous notion that house prices in the United States would rise perpetually. The IndyMac strategy was to originate individual mortgage loans, bundle them together into securities, sell these securities as quickly as possible to investors on Wall Street, and repeat ad infinitum. The strategy was jejunely simple. From IndyMac's inception as a savings association in July 2000, Perry grew the firm aggressively so that by 2006 it employed around 6,500 people and was the seventh largest savings association and ninth largest originator of mortgage loans in the entire United States. However, when the housing boom came to a sudden end, so did IndyMac. Its collapse in the summer of 2008 was one of the largest bank failures in American history. It cost the Deposit Insurance Fund many billions of dollars. Its chief risk officer—John DelPonti, appeared to do "all right." He had put in place a *sweetheart deal* for himself.

Too often senior leaders have driven their businesses through the gates of ultimate catastrophe. Tempted by acclaim, accolades, and wealth, they blindly pursue an unsustainable business strategy. And rather shamefully, they see the risk manager as their last resort, a put option that if exercised will enable them to continue uninterrupted onwards with their calamitous vision. Risk managers, they believe, *can eat their sins.* Unfortunately, many a risk manager is all too eager, whether through lack of understanding or dubious moral fiber, to perform this unsavory role. But having one's business sins eaten by one's risk manager is not a rescue, it is just an unorthodox

I'M ALRIGHT JACK!

Risk-Free Real Estate for the Chief Risk Officer?
March 28, 2005 at 10:20 A.M. by ML

Buying a home in today's overheated real estate market can be so risky as this recent article points out. And who needs that kind of stress? Certainly not a bank executive whose title is Chief Risk Officer. Thankfully, IndyMac (IMB) has come to the rescue of John DelPonti, the 40-year-old former PwC partner it hired last March. In the recent proxy, IndyMac notes that it purchased a $2.47 million home for DelPonti, which it is renting to him for $6,500 a month. For DelPonti, it's a sweetheart deal. The rent, which is structured under a five year month-to-month lease, is significantly below the $14,000 a month he'd have to pony up if he had to take out a mortgage like ordinary people do. Plus, he's protected no matter what happens to real estate prices in Pasadena. If prices rise, the agreement allows him to buy the house for the bank's purchase price. And, if they fall, he can buy it for whatever the current appraised price is. Even more surprising is that DelPonti isn't even one of the top five executives at the company, which kind of makes you wonder what sorts of real estate deals IndyMac dished out to the top cheese. (See Footnoted: http://www.footnoted.com.)

way of redeeming oneself, of attempting to gain something that one does not deserve. And ultimately, it will not work, even if one's risk manager is an eager and willing sin eater. Indeed, by all accounts Richard Munslow was a wonderful sin eater, a man who had mastered his craft. But on his demise in 1906, not a single soul came forward to dine off his corpse! He was eventually buried, unredeemed, under the sod in the ancient graveyard of St. Margaret's Church, Ratlinghope. And this, I told the young man, is the reason why you must understand, accept, and embrace the golden rule of risk management. Without pausing for the individual to ask the inevitable follow-up question, I dashed out of the conference hall, hailed a taxi, and caught, by a hare's breath, my train.

Odysseus and the Sirens' Song

What is the golden rule of risk management? You will not find it being much discussed at conferences or taught in courses on risk management. In many cases, the topics discussed in these forums are highly quantitative, requiring advanced knowledge of mathematics, statistics, and numerical methods. Or else the themes are legalistic with bone dry overemphasis on discipline of conduct. Yet in reality, neither the rules of mathematics nor knowledge of laws or regulations can advance your understanding of risk management very far if you fail to understand, accept, and embrace the golden rule of risk management. It can be explained by use of the allegory of Odysseus and the Sirens' song[7]:

> Homer's Odyssey has one of the earliest examples of solutions to the problem of management becoming enraptured with a course of action and becoming blind to the course's disastrous consequences. Odysseus' solution was to have his crew bind him to the mast and to put wax in their ears. These measures freed him to hear the song and enjoy it but left him unable to steer his vessel towards the Sirens and the rocks on which they sat. If Odysseus had not plugged his crew's ears, all would have enjoyed the Sirens' song and all would have been well until the last moment when the boat smashed upon the rocks. … If I may pursue the metaphor of the Sirens' Song a little further, it is interesting to note that Odysseus' solution had two parts. His arrangements ensured that he could hear but not steer, and that the crew could steer but not hear. Odysseus made sure that those who imposed the constraints, that is, tied him to the mast, and who could therefore untie him, were not subject to the same influences as he was. In our context of the management of firms, it is important that those who ultimately impose the rules not be responsive to the same influences as those to whom the rules apply.

That the King of Ithaca could conceive of such a strategy underscores his legendary guile and resourcefulness. Today, his approach seems so blatantly obvious that its implementation within the context of risk management, one might have thought, had occurred long ago. It has not.

The Consequence of Ignoring the Golden Rule

Enron ignored the golden rule. This severely undermined the ability of its risk managers to contribute effectively to a sustainable business strategy[8]:

> The firm maintained a risk management function staffed with capable employees. Lines of reporting were reasonably independent in theory, but less so in practice. The group's mark-to-market valuations were subject to adjustment by management. The group had few career risk managers. Enron maintained a fluid workforce. Employees were constantly on the lookout for their next internal transfer. Those who rotated through risk management were no different. A trader or structurer, whose deal

a risk manager scrutinized one day, might be in a position to offer that risk manager a new position the next. Astute risk managers were careful to not burn bridges. Even worse, risk managers were subject to Enron's "rank and yank" system of performance review. Under that system, anyone could contribute feedback on anyone, and the consequences of a bad review were draconian. Risk managers who blocked deals could expect to suffer in "rank and yank."

Intentional disregard of the golden rule strikes at the very ethos of risk management. By ethos of risk management, we mean, in part, risk management's ability to function as a place for rational impartial expression, incorporation of knowledge, and clarifying analysis of the level of risk relative to risk appetite.

It seems the critical import of the golden rule may not be clearly understood by executive leaders or even risk managers themselves. Take for example, a common situation that can arise with the chief financial officer (CFO). As the CFO's ultimate responsibility is financial reporting and financial management,[9] it might appear a quite natural state of affairs for him to assume the responsibility for risk management. Indeed, this is the situation that exists at many large corporations, including during the early 2000s, at the Federal National Mortgage Association (Fannie Mae).

However, having the CFO act as the head of risk management violates the golden rule of risk management. To some this may seem surprising. But after a moment's reflection, it will become obvious. Since many of the most serious risks facing a company are financial in nature, and therefore fall under the remit of the CFO, there is quite clearly a conflict of interest. The golden rule of risk management informs us that ultimate responsibility for financial functions and risk management should be separate. And this can be a difficult message for executive leaders and boards to accept. But the effective management of risk depends heavily on this principle.

Fannie Mae, during the 2000s, ignored the rule. It appointed its CFO as the chief risk officer (CRO). The consequences were perfectly predictable and absolutely inevitable. The Office of Federal Housing Oversight in a scathing report stated[10]:

> … we found that the Chief Financial Officer also serves as the Chief Risk Officer of Fannie Mae, and is directly responsible for overseeing the Enterprise's Treasury and Portfolio Management functions, in addition to the Controller's Department. *The combination of these responsibilities does not provide the independence necessary for an effective Chief Risk Officer function.* We further found that Mr. Howard was instrumental in setting financial targets as Vice Chairman, and had the authority to meet these targets as Chief Financial Officer. (Emphasis provided in report.)

The report further stated[11]:

As part of his responsibility for the retained portfolio, the CFO has the authority to approve transactions related to mortgage acquisitions and derivatives. He also has the authority to set risk management strategies which are used to develop the financial forecast. Additionally, he has the authority to determine how the financial transactions are reported in the financial statements. This lack of segregation of duties is inappropriate ...

An Immutable Condition for Success in Risk Management

There are lessons risk managers, senior executives, and boards must learn from Enron and Fannie Mae's failure to adhere to the golden rule. Without independence, risk management will yield to the Sirens and the rocks upon which they sit. And the entire business-wide risk management framework will have a deep structural weakness, which will eventually surface, probably with dire consequences, as Robert Jones, the chief risk officer of Amaranth Advisors LLC can attest. Jones failed to comprehend that for good reason in commercial banks, the risk management function is independent from the business function. This is critical in order to protect the integrity and objectivity of risk assessment, pricing, and management process. How else can we objectively ensure that risk undertaken is commensurate with both risk appetite and capacity to manage such risk?

Independence requires regular and accurate monitoring and reporting of the level of risk to ensure that even if there were a major shock, the business could sustain the loss incurred and continue to be a profitable entity. In other words, it is a critical element of business strategy and sustainability. Conflicts of interest are inevitable if monitoring and communicating the level of risk undertaken is combined with the business function of taking on the risk. Active and independent risk management, internal audit, and compliance functions are crucial elements of a bank's internal control process. One would have hoped something along these lines was rigorously enforced by a *sophisticated* hedge fund such as Amaranth; apparently, it was not.

Independence of the risk management function is an immutable condition for successful risk management. As Felix Kloman, long-time commentator on risk management issues, makes clear[12]:

> Independence of risk management is necessary to permit and stimulate both strategic perspective and the courage to speak out when required.

It is increasingly specified in best practice guidelines; for example, the Group of Thirty[13] clearly states that risk management must be fully independent of the risk-taking business. Individuals responsible for aspects of a risk review (risk managers, internal audit, compliance, and so on) must be independent from risk-taking business units and report directly to boards or senior management who are not involved in risk-taking activity. Independence also requires a separate oversight and reporting structure distinct from audit and financial issues. Glyn Holton, the long-standing editor of Risk Management Reports, outlined specific criteria against which to assess adherence to the rule[14]:

1. Risk managers have reporting lines that are independent from those of risk-taking functions.
2. Except at the highest levels, risk takers have no input on the performance reviews, compensation, or promotion of risk managers, and conversely.
3. Employees cannot switch from one role to the other. Those hired into risk management stay in risk management; those hired as risk takers stay as risk takers.
4. Risk managers do not take risks on the firm's behalf. They do not advise on which risks to take. They express no opinions about the desirability of any particular risks.

These criteria require as a minimum the development of appropriate risk assessment and management policies, methodologies, and procedures. These will increase the likelihood that risks are identified, evaluated, monitored, reported, and effectively mitigated. The scope of work and functions of an independent risk management function fall essentially into four groups:

1. Risk policy, methodology, and the risk information system.
2. Risk assessment.
3. Portfolio monitoring.
4. Review of problems and assistance in their workout.

The depth of the economic crisis arising out of the subprime mortgage crisis of 2007 have given many risk managers a deeper appreciation of the import of the golden rule. The crisis has served as a catalyst for change because it challenged the efficacy of an erroneous risk management model—a model practiced by hedge funds, energy companies, and even government-sponsored enterprises. The financial crisis of 2008, and 2008 revealed the importance of preemptive and independent risk management. Unfortunately, the golden rule does not appear much in the financial newspapers or on the late night business television shows. On occasion, an insightful commentator might bring it to the attention of a wider audience. But, it is soon forgotten, often disregarded by executives and those with the real power to ensure it is enforced. And this is a terrible mistake. The staggering losses to investors and the public arising from the failure of companies such as IndyMac and Amaranth Advisors provide a powerful

VALUE-ADDED KEY POINTS

Key Point 1: The golden rule of risk management underpins the ethos of risk management. By ethos of risk management, we mean, in part, risk management's ability to function as a place for rational impartial expression, incorporation of knowledge, and clarifying analysis of the level of risk relative to risk appetite.

Key Point 2: The golden rule provides a solid bedrock upon which to build a genuinely risk-aware culture and provides a basis for clear articulation and monitoring of a company's risk tolerance.

Key Point 3: The rule informs us that effective risk management requires a risk management function that is fully independent of the business units that generate risk exposures.

Key Point 4: The very real prospect of risk managers mutating into *risk sin eaters* provides a fundamental rational for independence. Independence enhances the likelihood of risk taking being fully aligned with risk appetite and diminishes the likelihood of the risk management function becoming bloated by compensatory *beer, bread, and salted meat*.

Key Point 5: Independent risk management provides an internal check against any incentives for individual units or employees within the firm to hide risk exposures from senior management.

Key Point 6: Without independence, risk management will yield to the Sirens and the rocks upon which they sit. The entire business-wide risk management framework will have a deep structural weakness, which will eventually surface, probably with dire consequences.

Key Point 7: Unfortunately the golden rule may not be clearly understood by executive leaders or even risk managers themselves.

Key Point 8: Neither the rules of mathematics nor knowledge of laws or regulations can advance your understanding of risk management very far if you fail to understand, accept, and embrace the golden rule of risk management.

testimony as to why you must understand, accept, and ultimately embrace the golden rule of risk management.

For Further Thought

Questions the reader might like to consider are given below.

1. Given the tremendous advances in financial risk measurement and management, why was the solvency of large and complex financial firms threatened in 2008 by large losses in the mortgage market?
2. How does your organization price risk?
    i.   What role does confidence play in the determination of the level of risk?
    ii. Who would tell you that risk taking is excessive?
    iii. What objective mechanism do you have in place to validate the level of risk taking?
3. What are the incentives in place to ensure you and your risk management team are aligned to think proactively about your risks?

Additional Resources

Martin (2010) and Golub and Crum (2009) discuss the role of the golden rule within the context of the 2007 financial crisis. They argue much of the economic crisis of 2007 to 2010 can be attributed to a failure of risk management processes across a variety of financial services firms. Holton (2004) looks at the failure of the risk management paradigm of the 1990s and explores the gap between the golden rule and risk management practice. The Office of Federal Housing Enterprise Oversight (2004) report provides additional details on the consequences of Fannie Mae breaching the golden rule. Kloman (2004) discusses why a change of attitude is necessary in order for the golden rule to find widespread acceptance. The Group of Thirty (2003) highlights a number of corporate financial management and reporting scandals which illustrate the golden rule. The report goes on to propose a set of best practices for governance and financial reporting. Tschoegl (1999) discusses a number of financial risk management failures within the context of the golden rule.

Golub, B.W. and Crum, C.C. (2009). Risk Management Lessons Worth Remembering from the Credit Crisis of 2007–2009. Abstract. *SSRN* October 31. http://ssrn.com/abstract=1508674.
Group of Thirty. (2003). *Enhancing Public Confidence in Financial Reporting*. Washington, DC: Group of Thirty.
Holton, G.A. (2004). A New Position on Risk. *Futures and Options World* (February): 44–45.
Kloman, F. (2004). Skills and Aptitudes Risk Management Reports. *Society of Actuaries: Risk Management Newsletter* (July) (2):12–14.
Martin, P. (2010). Why Is Operational Risk Management Important? *Journal of Securities Operations & Custody* 2(4) (January): 324–332.

Office of Federal Housing Enterprise Oversight. (2004). Report of Findings to Date Special Examination of Fannie Mae. *Office of Compliance.* http://www.ofheo. gov/media/pdf/FNMfindingstodate17sept04.pdf.

Tschoegl, A.E. (1999) *The Key to Risk Management: Management.* Wharton Financial Institutions Center Paper 99-42-B. Philadelphia, PA.

Endnotes

1. Matthew Moggridge gave an account of his encounter with a sin eater to the Cambrian Archaeological Association at their meeting in Ludlow in 1852: "When a person died, the friends sent for the sin-eater of the district, who on his arrival placed a plate of salt on the breast of the defunct, and upon the salt a piece of bread. He then muttered an incantation over the bread, which he finally ate, thereby eating up all the sins of the deceased … [The sin eater] vanished as quickly as possible from the general grave: for as it was believed that he really appropriated to his own use and behoof of the sins of all those over whom he performed the above ceremony, he was otterly detested in the neighbourhood—regarded as a mere Pariah—as one irredeemably lost." For further details, see *Encyclopedia of Religion and Ethics, Part 22,* 2003 ed., s.v. "sin-eating," by James Hastings and John A. Selbie, Elibron Classics Series, Edinburgh, p. 573.
2. See *Wall Street Journal* online.
3. See the *Post-Gazette* article, What Went Wrong at Amaranth Advisors, by Ann Davis, Henny Sender, and Gregor Zuckerman (http://www.post-gazette.com/).
4. But this is not totally surprising as discussed in Chapter 4.
5. See, for example, the *Washington Post*, From the Ex-Employees: Revenge, Shock, Sadness, by Frank Ahrens, Friday, May 26, 2006.
6. See the *New York Times*, Ex-Enron Officer Says He Warned of Shady Partnerships, by Alexei Barrionuevo, March 15, 2006.
7. See Tschoegl (1999).
8. See Holton (2004).
9. In particular ensuring a strong balance sheet and understanding the inherent volatility in reported earnings.
10. See Office of Federal Housing Enterprise Oversight (2004).
11. Violation of the golden rule of risk management is not the only argument against having the CFO take on chief risk officer responsibilities. Awareness of risk in its many different forms has grown dramatically over the past decade or so. Today risk management is much more expansive and engaging it covers much more than traditional financial risks.
12. See Kloman (2004).
13. See Group of Thirty (2003). The Group of Thirty, established in 1978, is a private, nonprofit, international body composed of senior representatives of the private and public sectors and academia. It aims to deepen understanding of international economic and financial issues. Details of its publications and current activity can be found at Group of Thirty (www.group30.org/).
14. Published at Riskinfo (http://www.riskinfo.com).

6

A Powerful Secret from Henry Fayol

Few people would be surprised to learn that, as a rule, we most prefer to focus on the return from a promising deal rather than the risk of possible losses. The thought of making 10%, 50% or more and the consequent pleasure in consuming those winnings implants contented thoughts in our mind. Researchers have been investigating this and many other behavioral biases since the 1950s.[1] In 1962, Robert Kates, a geographer working in Chicago, observed that people refuse to buy flood insurance even when it is heavily subsidized and priced far below fair value.[2] This is bizarre behavior indeed and anathema to economics notion of a rational man.

A flurry of academic research aimed at understanding this seemingly irrational behavior was initiated by the environmental hazard community. It was soon discovered the construction of dams and levees, which are designed to reduce the frequency of floods, resulted in a large proportion of individuals refusing to purchase flood insurance.[3] This appears on first blush very irrational behavior. This is partly because after construction of a levee or dam, the cost of flood insurance generally falls, if anything one might expect homeowners and businesses to increase their coverage. It also appears curious because while dams and levees decrease the frequency of floods, damage per flood is much greater. This was seen in New Orleans during 2005 when Hurricane Katrina poured a tidal wave of water smashing through the city's levees. The damage was catastrophic. Seven years later in 2012, the city had not fully recovered. It appears the presence of a dam or levee creates a false sense of security, leading to a reduced desire to purchase flood insurance protection. Even though when a flood occurs, it is much more likely to be catastrophic—precisely the circumstance insurance is designed to mitigate!

Erroneous judgmental biases appear to exist in many areas of human activity, including in our thinking about death. Sarah Lichtenstein, a decision research scientist working in Eugene, Oregon, addressed the question, "How do people judge the likelihood of death from various causes?"[4] Alongside a number of colleagues, an experiment was devised to answer four specific questions:

1. How well can people estimate the frequencies of the lethal events they may encounter in life?
2. How small a difference in frequency can be reliably detected?

3. Do people have a consistent internal scale for such events?
4. What factors, besides actual frequency, influence people's judgments?

As expected, the researchers found individuals tend to have a good general grasp of which risks cause the largest numbers of deaths and which cause the fewest deaths. However, when asked to quantify risks numerically, they found people severely overestimate the frequency of rare causes of death, and severely underestimate the frequency of common causes of death. Lichtenstein and her colleagues also discovered a tendency for individuals to exaggerate the frequency of certain specific causes of death and to underestimate the frequency of others: accidents were judged to cause as many deaths as disease despite the fact that diseases cause over 10 times as many deaths as accidents. Homicide was incorrectly judged a more frequent cause of death than, say, diabetes or stomach cancer. The authors explain:

> Events that capture our attention and "stick in our mind," like homicide, may appear more frequent than they are. Rare events may be overestimated because their appearances are well spread and distinct. Catastrophic (multi-fatality) events may be overestimated because of their salience or underestimated because of massed presentation. ... Thus we might expect that the frequencies of dramatic events such as cancer, homicide, or multiple-death catastrophes, which tend to be publicized disproportionately, would be overestimated, while the frequencies of "quiet killers" would be underestimated.

A follow-up study[5] looked at the reporting of deaths in two newspapers. It found underestimation and overestimation errors were highly correlated with reporting in newsprint. The correlation was strongest with risks in which there was considerable scientific uncertainty. This was a strange finding. It led Kumanan Wilson, a Canadian researcher, to investigate the role the media plays in communicating scientific information to the public and policy makers.[6] The study looked at how the Canadian print media reported the theoretical risk of blood transmission of Creutzfeldt-Jakob disease (CJD). CJD gained a high degree of public attention in many Western countries because it was the first major infectious challenge to the blood supply after human immunodeficiency virus (HIV) and hepatitis C. Careful analysis of 245 newsprint articles led Wilson and his colleagues to conclude Canadian media can influence the public perceptions of the severity of a risk.

Disproportionate reporting of certain types of risk appears to influence our perception of that risk. If you read any popular newspaper, you may have observed a tendency to dwell on the potential for catastrophe in some industry or other rather than the day-to-day (and less exciting) success. Almost every successful chief investment officer will have fought those who have held this bias. This is especially the case in relation to hedge funds. Despite being of lower risk than the general stock market, the widely held perception by boards, trustees, and the public is otherwise. Bestselling author Richard Bach, in his book *Nothing by Chance: The American Way*, captures the inevitable consequences of this bias in a passage about the irrational fear shown by a young married couple from Wisconsin traveling on their very first flight:

> In all that wind and engine blast and earth tilting and going small below us, I watched my Wisconsin lad and his girl, to see them change. Despite their laughter, they had been afraid of the airplane. Their knowledge of flight came from newspaper headlines, a knowledge of collisions and crashes and fatalities. They had never read a single report of a little airplane taking off, flying through the air, and landing again safely. They could only believe that this must be possible, in spite of all the newspapers, and on that belief, they staked their three dollars and their lives. And now they shouted and smiled to each other, looking down, pointing.

What might be startling to note is that these individual behavioral biases persist at the corporate level, even in those institutions whose entire business revolves around quantifying, managing, and repackaging of risk. The clearest illustration I know relates the U.S. residential housing bubble of the early 2000s and a behavioral bias known as an availability cascade.[7] An availability cascade is a self-reinforcing process of collective belief formation. It begins with a strongly held belief, which in turn triggers a chain reaction that gives the belief increasing plausibility. Confidence in the belief grows as it gains traction in the public mind via television, radio, newspapers, social networking sites, and word-of-mouth.

During the late 1990s through the mid-2000s, the Clinton and then the Bush administration encouraged home ownership by urging banks to relax lending standards. Home ownership became an essential part of the so-called American Dream. Fanned by a gaggle of reality television shows, radio infomercials, and newspaper real estate *rags to riches* articles, the belief that house prices could only go up gradually took hold. Subprime mortgage borrowers, many barely eking out a living, bet their futures on a housing bubble they believed would never burst. Many U.S. banks, some very large, over time acquired substantial exposure to subprime mortgage debt on their balance sheets. These extremely risky assets were primarily financed by short-term borrowing. The collapse in U.S. residential house prices beginning in 2006 resulted in the closure of the market for these residential mortgage securities. Estimated losses at one point topped $250 billion. Liquidity dried up. The market for interbank funds albeit vanished resulting in an economy-wide sharp contraction in credit availability. The U.S. economy plunged into the deepest recession since the Great Depression. The crisis extended to the global economy, throwing much of the developed world into a deep financial and economic malaise.

What is interesting is that prior to the crisis all the major warning signs were present. Over the period 2000 to 2006, house prices in the metropolitan areas of Tampa, Miami, San Diego, Los Angeles, Las Vegas, and Phoenix grew at a rate of more than 10% per year. Clearly unsustainable. Over the years 1990 to 2000, U.S. household debt had grown at a modest rate of 1.2% per year. It exploded to a rate of increase of 4.2% per year from 2000 to 2006. Real economic growth in the U.S. averaged a paltry 1.9% over the first decade of the 21st century. This was considerably less than the historical average of around 3.9%. Simple signs. But missed or largely ignored by some of the most prestigious financial institutions in the world. Why did such a widespread miscalculation of risk by large sophisticated financial firms occur? A portion of the explanation lies in behavioral biases exhibited at the corporate level. Part of the solution lies in Henri Fayol's notion of a *strategic security director*.

The Great Work: General and Industrial Management

Toward the end of the 19th century, the Frenchman Henri Fayol,[8] perhaps the greatest management theorist of his day, hinted at the idea of senior executive involvement in corporate risk management. His great work, *General and Industrial Management*, was the first systematic analysis of management practice within the context of a theoretical and scientific framework. The very first English translation of his work was printed by the International Management Institute in Geneva. Only a few hundred copies were made available to Sir Isaac Pitman & Sons, Ltd., for distribution in the United Kingdom. No English translation was published in the United States of America despite a great deal of interest in management theory.[9] English-reading audiences did not have widespread access to his ideas until almost four decades after the initial publication[10]:

> He [Henri Fayol] identified clearly the "strategic security director," the ancestor of the CRO. It took 40 years for Fayol to be translated into English, 40 more years for the Americans to read him.

Many of Fayol's ideas found a foothold in the broad discipline of management science, but somehow risk management, in particular, the idea of a *strategic security director*, was for the most part overlooked. Part of the explanation may lie in the observation[11]:

> There is nothing inherent in the basic steps of risk management that dictate where it should be placed on the organizational chart. Many other disciplines, both academic and practical, such as marketing, accounting, etc., have well established patterns and positions on most firm's organization charts that are often separate and distinct from every other discipline. This helps give each of these disciplines a raison d'être, and internal and external validation. This is not the case with risk management. There are few external risk management firms which undertake

solely risk management, and which are as respected by senior management as a large accounting firm or marketing agency.

A more telling explanation is that the strategic security director challenges the traditional executive structure. Chief finance officers (CFOs), treasurers, and line managers may be resentful of a new third-party C-level executive. Their inevitable concern over a perceived loss of power may make them somewhat leery of the position. Chief finance officers, in particular, being the drivers of the budget and financial planning and *de facto* risk managers as part of their financial management function, may feel more resentful than most. They may incorrectly perceive Fayol's strategic security director as simply themselves, treasurer, line managers (or some other combination)—risk management is their responsibility. But the golden rule of risk management informs us that ultimate responsibility for financial functions and risk management should be separated.

The reality is that CFOs, treasurers, and line managers have not traditionally looked at risk through a single lens. Risk overseers in different business units have tended to manage the risks of foreign exchange, interest rates, commodities, and insurance with very little, if any cross communication. This traditional approach to risk management has resulted in individual risk silos. For example, technology risks such as Internet security may be handled by the information technology department; hazard risks by the corporate risk department, and capital acquisition and market risks by the CFO. Excessive risk concentrations across business units have not necessarily been identified, while natural hedges have not been exploited.[12]

Henri Fayol's notion of a strategic security director seems more relevant today than ever—a senior level individual who signals senior managements' willingness to take responsibility for their risk management and, in particular, ensure risk management activities from diverse business units come together effectively to allow comprehensive assessment of risk. The chief risk officer (CRO) should be a thought leader in developing risk management programs, processes, and policies—and not a threat to the hegemony of the CFO or other senior executives. Today, risk management is much more expansive and engaging. It covers much more than traditional financial risks. Corporations need to assess risks accurately in order to mobilize their resources efficiently. The scope of the risk management task and consequences of failure demand, as Henri Fayol correctly identified over 60 years ago, a dedicated senior level risk position.

The Rise of Fayol's "Strategic Security Director"

As recently as a decade or so ago, corporate Europe, America, and Japan combined had only a handful of chief risk officers, almost exclusively working within large banks. This lack of numbers was especially puzzling in the case of the financial services industry given that a key feature of their value creation lies in their ability to allocate risk efficiently through the trading, bundling, and unbundling of the risks of various financial contracts. Until very recently, the necessity for senior management of small- and medium-sized financial corporations to allocate resources toward setting up a corporate-wide risk management framework was absent. It was perceived as an unnecessary expense.[13] This view held sway despite the fact that financial service firms large and small originate, trade, and service financial assets; indeed their core business is to transform, manage, and underwrite risk.

The situation only began to really change, with risk management slowly emerging as a recognized professional discipline,[14] following a decade of financial fiascos ranging from the failure of Barings, the Enron scandal, the WorldCom crisis, the collapse of Bear Stearns, the titanic implosion of Lehman Brothers, right through to the uproar at that most respectable of institutions, Fannie Mae.[15] In the light of these and other allegations of failures in corporate risk management, politicians, regulators, professional associations, academics, and even companies themselves have responded by demanding a more comprehensive approach to the mechanisms through which companies are directed and controlled. Part of this trend has been a growing recognition that today's business environment requires robust and effective risk management. Today, the chief risk officer increasingly sits at the helm of this activity.

A growing number of North American, European, and Asian corporations are gradually recognizing the necessity of including risk management at the very highest level of their operations via an identifiable CRO who acts to provide risk assessment, risk management, and risk assurance.[16] One of the first steps Allied Irish Bank took in the wake of its 2002 troubles brought on by a rogue trader was to hire a CRO.[17] In the spring of the same year, following the collapse of Enron, players in the U.S. energy sector formed the Committee of Chief Risk Officers. Today, it consists of more than 30 of the leading actors in the energy sector.[18] A survey by consulting firm Deloitte & Touche found the number of CROs grew 65% between 2002 and 2005 in the financial services sector.[19] A separate survey found that around 45% of major companies in the U.S., Europe, and Asia had in place a CRO on or before autumn 2005, another 24% intended to have one in place within a few years.[20] Thus, the idea of the CRO, although not new, has grown in importance particularly over recent years. Part of the impetus has come directly from regulators—one of the components of the settlement between the secondary mortgage market giant Fannie Mae and U.S. federal regulators was the appointment of a CRO.[21]

The recent financial and economic turmoil has accelerated the trend. In June 2010, Legal and General, founded in 1836 in one of Chancery Lane's famous coffee shops, appointed its first chief risk officer. In the same year, Torus, the global insurer, appointed a group chief risk officer and the Saudi

British Bank followed suit, creating a new post of chief risk officer.[22] While the trend is unmistakable, what is less clear is the appropriate scope of the CRO's responsibilities.

I suspect there are many similarities between the state of knowledge of management science when Henri Fayol produced his pioneering work and our understanding of the scope of responsibilities of the CRO today. Fayol created a theoretical framework in order to assess and enhance the efficiency of the very practical business of management. Unfortunately, such a rigorous theoretical framework applied to the thought leader on risk management has yet to be fully developed. Obscurity over the scope of CRO responsibilities remains for many practitioners a barrier to their acceptance of the validity of the role. In many ways, this reticence is somewhat of a disappointment. A disappointment because there is a well-established principle, which if understood and applied, can serve as a solid foundation upon which to build an effective suite of responsibilities for the CRO. What is this principle? In modern times, it has become known as the Warren Buffet principle of risk management.

The Warren Buffet Principle of Risk Management

Suppose that while leafing through the newspaper, you notice an advertisement for volunteers to join the board of directors of a very large and prestigious financial organization. Let's suppose further that, finding the idea intriguing, you contact the address in the advertisement, and are invited to an interview. When you arrive at the interview, a location in the business district of town, you are led into an oak paneled room lined with paintings of steely-eyed, bearded gentlemen, all wearing dark suits. Six existing board members, all suited, bearded, and steely-eyed, sit opposite you and the interview begins. Of course, you are nervous; butterflies spur you on to give the best performance of your life! You are offered the position, which comes with a substantial honorarium, one so large it dwarfs your regular paycheck. What luck!

Your very first task (after buying a new suit, growing a beard, and perfecting your steely-eyed look in the bathroom mirror) is to vote on the appointment of a CRO, one recommended by the chief executive officer (CEO) and executive team. You vote in favor and the CRO is approved. The CEO then delegates key risk management tasks to the CRO, who is expected to report to the board on issues regarding enterprise-wide risk on a regular basis. The situation runs smoothly for a number of years, the honorarium payments mount up, and you are very happy. Of course, no good deed goes unpunished, and you soon find yourself, along with the other board members, caught up in a financial calamity, a crisis so large, that several of your competitors fail.

Panic grips the entire financial system. Your organization, which has taken on substantially more risk than the board was led to believe, is advised to ask the government for emergency funds. Without government assistance, your firm (and your honorarium) will go the way of your bankrupt rivals. Your firm asks for and receives government assistance. The media, like a pack of wild dogs, pounce on perceived weakness of the board and executive team. Following months of stinging and barely factual newspaper editorials, against a backdrop of angry public protests and political finger-pointing, an emergency board meeting is convened. The chief executive officer demands the resignation of the CRO. The argument is clear—the CRO failed to advise the CEO and board on the level of risk the company was exposed to. The board agrees and the CRO is fired.

This is a plausible scenario, which no doubt has been played out many times over in recent times. However, placing risk control exclusively in the hands of the CRO violates the Warren Buffet principle of risk management[23]:

> In my view, a board of directors of a huge financial institution is derelict if it does not insist that its CEO bear full responsibility for risk control. If he's incapable of handling that job, he should look for other employment. And if he fails at it—with the government thereupon required to step in with funds or guarantees—the financial consequences for him and his board should be severe.

The principle provides a reference for risk taking and risk appetite. It articulates a fundamental rule of risk management—the CRO is not the ultimate manager of risk; that responsibility lies with the CEO. Neither is the CRO a put option to be exercised by the CEO in difficult times to protect his own position. Failure in the management of risk is a failure in executive leadership.

It appears too many CROs languish underutilized and alone in the back rooms of the corporations for which they work. There they may be found fiddling with their quantitative models or else muttering incoherently about tweaks in their measure of risk for illiquid assets. Of course, none of this has a scintilla of relevance to the rainmakers. They go about their business as usual making money and generally avoiding the CRO or his minions. On occasion, as did Enron or Amaranth Advisors, the CRO is brought out into public view, primarily to impress a specific constituency. Then, they are cast back into the corporate bowels, that is, until a disaster strikes. When it does strike, as board members and executive officers scream "off with his head," the dazed risk officer may find himself thrown as a sacrifice onto the altar of public, political, or market opinion—a bloody sacrifice to appease the gods of business. Thoughtful, ambitious, self-motivated managers understand the inherent dangers of an officership in risk management and may avoid it altogether on their climb up the corporate ladder[24]—and who can blame them?

It is well known that officerships like this can be a dumping ground for problems, or a way of telling the world that one is serious about an issue. In this respect CROs may be an "organisational fix," and a risky position as a potential blamee.

Can Chief Risk Officers Add Value?

The evidence is now only beginning to be documented. Researchers Liebenberg and Hoyt[25] found that firms with greater financial leverage are more likely to appoint a CRO. Scholars Pundmann and Kobel observe[26]:

> Most businesses want to minimize their liability and related management costs, and are betting that a good enterprise risk-management plan, integrated with a chief risk officer (CRO), will make that happen.

Broader empirical evidence supporting the value of a CRO is also beginning to emerge[27]:

> The research indicates that top risk managers now play a central role in coordinating their firm's response to an unprecedented range of threats. The main benefit of appointing a CRO, according to 52% of executives in the survey, is that they can expand risk management to address more risks. They also enable the business to make better investment decisions, in particular by bringing a more effective approach to measuring and comparing risk and reward.

To operate as an effective brake requires the CRO to be an active participant in the process that delivers sustainable profitability to the firm. This entails the CRO play a key role in four core areas:

1. Ensuring risk management discipline.
2. Articulating the desired risk profile of the company.
3. Imposing a common language when talking about risk.
4. Assisting the company in understanding those risks it wants to take on, those it wants to mitigate, and the tolerances and limits around those risks.

The position therefore has a leading role to play in crafting an organizational structure in which roles and responsibilities of individuals involved in risk taking are clearly defined and managed. This requires the full support and backing of the CEO. The CEO is expected to play center stage in setting the overall tone of corporate risk culture, one in which risk management is focused on improving the effectiveness of business processes by being built into corporate governance, business structures, planning and oversight of operational processes.

As a thought leader on risk, the CRO provides vision, passion, independence, and leadership. It carries a broad responsibility for[28]:

> Providing the overall leadership, vision, and direction for enterprise wide risk management; Establishing an integrated risk management framework for all aspects of risks across an organization; Developing risk management policies, including the quantification of management's risk appetite through specific risk limits; Implementing a set of risk metrics and reports, including losses and incidents, key risk exposures, and early warning indicators; Allocating economic capital to business activities based on risk, and optimizing the company's risk portfolio through business activities and risk transfer strategies; Improving a company's risk management readiness through communication and training programs, risk-based performance measurement and incentives, and other management programs; Developing the analytical, systems, and data management capabilities to support the risk management program.

What then are the most basic areas of accountability? First, to ensure material risks facing the firm have been identified. Second, ensuring the firm has in place an adequate mechanism to measure and model those risks. In other words, the CRO represents the risk management system and bears some responsibility for its design and oversight[29]:

> To make sure risk is properly understood and translated into meaningful business requirements, objectives, and metrics. Risks and rewards are clearly established to ensure that the corporate-liability view is pushed to the businesses and they're fully engaged in managing their portion of the risk.

Third, to ensure comprehensive monitoring of risks, and finally, overseeing, on the one hand the interpretation of risk, and on the other hand, communication of the nature of identified risks to senior management and the board in a timely fashion.

Thus, the CRO is an advocate for clarity and communication of the risks facing the firm. While the golden rule dictates the CRO will not have any operational responsibility, they must be in a position to oversee and influence policy over the entire risk management process. This includes, but is not limited to formalizing a risk charter, developing risk management policy and performance blueprint of existing and future risk management activities, and devising a risk management implementation strategy. Once these policies are in place, risk tolerances have been agreed upon, mitigation strategies put in place, and the appropriate risk profile determined, it is the business unit's responsibility, overseen by the CEO or board, to implement the risk management strategy or program. Used in this way, Henry Fayol's

VALUE-ADDED KEY POINTS

Key Point 1: Why does miscalculation of risk by large sophisticated financial firms occur? A portion of the explanation lies in behavioral biases exhibited at the corporate level.

Key Point 2: The notion of a CRO is not a new concept. Henri Fayol identified the strategic security director, the ancestor of the CRO in the early part of the 20th century.

Key Point 3: The CRO should be seen as the thought leader in developing risk management programs, processes, and policies rather than a threat to the hegemony of other C-suite executives.

Key Point 4: Risk management is much more expansive today than it was ten years ago. It covers much more than traditional financial risks. Its scope and the consequences of failure demand a dedicated senior level position—the CRO.

Key Point 5: The CRO, at a very minimum, should have some responsibility for:
1. Ensuring risk management discipline.
2. Articulating the desired risk profile of the company.
3. Imposing a common language when talking about risk.
4. Assisting the company in understanding those risks it wants to take on, those it wants to mitigate, and the tolerances and limits around those risks.
5. And playing a key role in crafting an organizational structure defining clearly roles and responsibilities of individuals involved in risk taking as well as managing it.

Key Point 6: To be effective, the CRO needs sufficient resources. Operational support for the CRO should come directly from a risk management group, the primary purpose of which is to build the capacity needed to address risk issues in a timely and efficient manner.

notion of a strategic security director may help to diminish the impact of those irrational behaviors first observed in the 1960s by a little known geographer by the name of Robert Kates.

For Further Thought

If you ask different companies what their CRO's mandate is, the variety of answers will be quite surprising. This is because, in practice, the mandate depends on the organization of the company, the company's business model, and the nature of support the CRO receives from the chief executive officer and the board. A value-added CRO is more than an administrator of compliance to risk management guidelines. He or she is a champion of risk management practices and culture across an entire company. Value-added comes from their ability to construct a risk governance structure that improves management decision making and maximizes return per unit of risk. Issues for further consideration include:

- Who are the drivers of risk management in your organization?
- What is the scope of the CRO mandate in your company?
- What are your greatest challenges in empowering the CRO?
- What steps has management taken to better understand the key risks of the company and empower the risk management function?
- What would you say is the value to your organization of the CRO? How do you know? What can be done to improve the value?
- What value-added would you see if the total resources dedicated to risk management were doubled?

Additional Resources

Kates (1962), Burton, Kates, and White (1978), Lichtenstein et al. (1978), Combs and Slovic (1979), Kahneman, Slovic, and Tversky (1982), Johnson and Tversky (1983), Shanteau (1987), and Wilson et al. (2004) discuss various aspects of behavioral biases and the influence of news media on the perception of risk. Fayol (1949), Lam (2000), Lamser and Helland (2000), Louisot (2004), Economist Intelligence Unit (2005), and Lee and Shimpi (2005) provide a historical perspective on the CRO. The Conference Board of Canada (2001) discusses how the position of CRO is evolving and where it is heading. Aabo, Fraser, and Simkins (2005) and Shaw (2005) trace out some of the consequences of risk management silos. Power (2004) touches on the role of the CRO. Independence of the risk management function is discussed in Tschoegl (1999), Group of Thirty (2003), Holton (2004), and Kloman (2004). The value-added of the CRO and risk management is touched on in Miccolis, Hively, and Merkley (2001), Liebenberg and Hoyt (2003), Pundmann and Kobel (2003), Carpenter (2004), and Corbett (2004). The Office of Federal Housing Enterprise Oversight (2004) illustrates the consequences of breaking the golden rule of risk management.

Aabo, T., Fraser, J, and Simkins, B.J. (2005). The Rise and Evolution of the Chief Risk Officer: Enterprise Risk Management at Hydro One. *Journal of Applied Corporate Finance* 17(3)(June).

Burton, I., Kates, R.W., and White, G.F. (1978). *The Environment as Hazard*. New York: Oxford University Press.

Combs, B. and Slovic, P. (1979). Causes of Death: Biased Newspaper Coverage and Biased Judgments. *Journalism Quarterly* 56:837–843.

Conference Board of Canada. (2001). *A Composite Sketch of a Chief Risk Officer*. Conference Board of Canada. http://www.conferencboard.ca/e-Library/abstract.aspx?did=1210.

Corbett, R. (2004). A View of the Future of Risk Management. *Risk Management: An International Journal* 6(3):35–50.

Economist Intelligence Unit. (2005). *The Evolving Role of the CRO: A New Report*. Economist Intelligence Unit: London. www.eiu.com/cro.

Edwards, W. (1954). The Theory of Decision Making. *Psychological Bulletin* 51.

Fayol, H. (1949). *General and Industrial Management*. London: Pitman & Sons.

Group of Thirty. (2003). *Enhancing Public Confidence in Financial Reporting*. Washington, DC: Group of Thirty.

Holton, G.A. (2004). A New Position on Risk. *Futures and Options World* (February):44–45.

Johnson, E. and Tversky, A. (1983). Affect, Generalization, and Perception of Risk. *Journal of Personality and Social Psychology* 45(1):20–31.

Kahneman, D., Slovic, P., and Tversky, A. (1982). *Judgment under Uncertainty: Heuristics and Biases*. Cambridge: Cambridge University Press.

Kates, R. (1962). *Hazard and Choice Perception in Flood Plain Management*, Research Paper No. 78. Chicago: University of Chicago, Department of Geography.

Kloman, F. (2004). Risk Management: Skills and Aptitudes. *Risk Management Reports* 31(5).

Lam, J. (2000). Enterprise Wide Risk Management and the Role of the Chief Risk Officer. *ERisk*. http://www.riskmania.com/pdsdata/Enterprise-wide%20Risk%20Management%20and%20the%20Role%20Chief%20Risk%20Officer-erisk.pdf.

Lamser, I. and Helland, E. (2000). How the Chief Risk Officer Is Taking Centre Stage. *Balance Sheet* 8(6) (June):26–28(3).

Lee, C.R. and Shimpi, P. (2005). The Chief Risk Officer: What Does It Look Like and How Do You Get There? *Risk Management* (September). The Risk and Insurance Management Society.

Lichtenstein, S., Slovic, P., Fischhoff, B., Layman, M., and Combs, B. (1978). Judged Frequency of Lethal Events. *Journal of Experimental Psychology: Human Learning and Memory* 4(6) (November):551–78.

Liebenberg, A.P. and Hoyt, R.E. (2003). The Determinants of Enterprise Risk Management: Evidence from the Appointment of Chief Risk Officers. *Risk Management & Insurance Review* 6.

Louisot, J.P. (2004). Managing Intangible Asset Risks: Reputation and Strategic Redeployment Planning. *Risk Management: An International Journal* 6(3):35–50.

Miccolis, J., Hively, K., and Merkley, B. (2001). *Enterprise Risk Management: Trends and Emerging Practices*. Altamonte Springs, FL: The Institute of Internal Auditor Research Foundation.

Office of Federal Housing Enterprise Oversight. (2004). *Report of Findings to Date: Special Examination of Fannie Mae*. Office of Compliance. http://www.ofheo.gov/media/pdf/FNMfindingstodate17sept04.pdf.

Power, M. (2004). *The Risk Management of Everything: Rethinking the Politics of Uncertainty*. London: Demos. http://www.demos.co.uk/files/riskmanagementofeverything.pdf.

Pundmann, S. and Kobel, B. (2003). Send in the Chief Risk Officer. *Optimize* 23(September).

Shanteau, J. (1987). Psychological Characteristics of Expert Decision Makers. In *Expert Judgment and Expert Systems*, Mumpower, J.L., Renn, O., Phillips, L.D., and Uppuluri V.R.R. (eds.). Berlin: Springer-Verlag.

Shaw, J. (2005). Managing All of Your Enterprise's Risks. *Risk Management Magazine* (September). Risk and Insurance Management Society.

Slovic, P. and Lichtenstein, S. (1971). Comparison of Bayesian and Regression Approaches to the Study of Information Processing in Judgment. *Organizational Behavior and Human Performance* 6:649–744.

Tippins, S.C. (2004). Risk Management: Where Is It and Where Does It Belong? *Risk Management: An International Journal* 6(3):35–50.

Tschoegl, A.E. (1999). *The Key to Risk Management: Management*. Wharton Financial Institutions Center Paper 99-42-B. http://fic.wharton.upenn.edu/fic/papers.html.

Urwick, L. Foreword in *General and Industrial Management*, Fayol, H. (1949). London: Sir Isaac Pitman and Sons.

Wilson, K., Code, C., Dornan, C., Ahmad, N., Hébert, P., and Graham, I. (2004). The Reporting of Theoretical Health Risks by the Media: Canadian Newspaper Reporting of Potential Blood Transmission of Creutzfeldt-Jakob Disease. *BMC Public Health* 4:1.

Endnotes

1. See, for example, Edwards (1954) who investigates how people make choices between gambles, Slovic and Lichtenstein (1971) discuss probability revision and probability learning, Kahneman, Slovic, and Tversky (1982) develop the notion of risk heuristics (mental rules of thumb) and biases, and Shanteau (1987) who investigates how the framing/context influences comparisons between experts and novices.
2. See Kates (1962). In particular, the study found people misjudge the hazard associated with floods. However, more frequent hazards are judged more accurately. Accuracy is also increased by both the recency of the hazard's last major occurrence and its impact on one's livelihood.
3. See Burton, Kates, and White (1978).
4. See Lichtenstein et al. (1978).
5. See Combs and Slovic (1979).
6. See Wilson et al. (2004).

7. Identified by Professors Kurian and Sunstein from Duke and Harvard Universities, respectively.
8. His work was initially published in French in 1916 and in English in 1949. See Fayol (1949).
9. See, for example, Urwick (1949).
10. See Louisot (2004). It is interesting to note when Henri Fayol was writing, the field of risk management as we know it today did not exist. However, the underlying concepts were there, and many, it not all of them were being applied in some way.
11. See Tippin (2004).
12. See, for example, Aabo, Fraser, and Simkins (2005) and Shaw (2005).
13. See, for example, Lee and Shimpi (2005).
14. And a discipline distinct from traditional insurance-based risk management.
15. Who ran afoul of its regulator amid accusations that it had violated accounting rules.
16. This is partly because as well as protection of the value of the business and its reputation, enhanced risk management is perceived as leading to the optimization of operational efficiency and improvement of decision-making processes. See Miccolis, Hively, and Merkley (2001).
17. A case of fraud at its Baltimore-based subsidiary, Allfirst, carried out by a currency trader named John Rusnak. Total losses to AIB were in excess of U.S. $750 million.
18. See Committee of Chief Risk Officers (http://www.ccro.org).
19. See surveys and research reports at Deloitte (www.deloitte.com).
20. See Economist Intelligence Unit (2005).
21. Adolfo Marzol was appointed the interim chief risk officer. He was replaced in November 2005 by Mark Winer, who was appointed Fannie Mae's deputy and acting chief risk officer.
22. See Arab News (http://arabnews.com/).
23. Quote by Warren Buffet and taken from the Berkshire Hathaway 2009 annual report.
24. See Power (2004).
25. See Liebenberg and Hoyt (2003).
26. See Pundmann and Kobel (2003).
27. Quote taken from press release from the Economist Intelligence Unit report into the role of the CRO. See Economist Intelligence Unit (2005) for full details of the report.
28. See Lam (2000).
29. See Pundmann and Kobel (2003).

7

The Incredible Advantage of a Monocle on Risk

Risk; solid, substantial, flashing red on a trader's screen, red as ash on Mars, or lava expelled from Mount Vesuvius. In the heat of the battle, you cannot look through it, nor yet gaze up and down it, nor over it. It seems impenetrable. When it erupts, it does so in a terrifying form; crushing the senses, untrustworthy, they lie addled, impotent, useless[1]:

> As this crisis climaxes, formerly reasonable people will start to predict the end of the world, armed with plenty of terrifying and accurate data that will serve to reinforce the wisdom of your caution. Every decline will enhance the beauty of cash until "terminal paralysis" sets in. Those who were over invested will be catatonic and just sit and pray. Those few who look brilliant, oozing cash, will not want to easily give up their brilliance. So almost everyone is watching and waiting with their inertia beginning to set like concrete.

Always, for some, risk's wake brings nightmare, mercilessly stripped of their wealth, tossed like somnambulistic souls naked into the antechamber to Hades, lackeyed by vanished dreams and now, too late, fully aware of their irretrievable mistakes. For others, who have learned how to carefully embrace it, risk brings joy, enrichment, and renewed life; yet in others, when it is passed, the crisis over, they only thought they saw it. Uninformed souls, who without learning or guidance, must become its next merciless victims. It is to this challenge which modern risk management must, in part, rise. Education and diffusion of risk management knowledge and principles throughout an organization is the inherent mission. An aspiration shared by all risk professionals and captured wonderfully by Dahl[2] in his uplifting short story about a house of ill-repute run by Madame Rosette:

> There was the smell of Cairo, which is not like the smell of any other city. It comes not from any one thing or from any one place; it comes from everything everywhere; from the gutters and the sidewalks, from the houses and the shops and the things in the shops and the horses in the streets and from the drains; it comes from the people and the way the sun bears down upon the people and from the way the sun bears down upon the gutters and the drains and the horses and the food and the refuse in the streets. It is a rare pungent smell, like something which is

sweet and rotting and hot and salty and bitter all at the same time, and
it is never absent, even in the cool of the early morning.

And, so it should be with risk management; its aroma must be ever present, penetrating all corners and levels of a corporation; every transaction,
deal, or new product must be bathed in its pungent aura. Risk is an inherent
part of any business and directors are invariably going to have to take it on in
pursuit of profitability; therefore, every director and every executive, indeed
all staff and decision makers, must be evaluating their business decisions
by reference to risk management principles. Key to infusing a *risk scent* into
the cultural fabric of an organization is the notion of a monocle on risk—an
integrated risk management framework,[3] the subject of this chapter.

What Is a Monocle on Risk?

It is helpful to begin with a simple definition. A framework is defined as an
open structure that gives shape and support to something; in our case that something is risk management. *Integration* refers to the aggregation of all risks
faced by a company and the strategic combination of risk management techniques to manage that risk. Thus, an integrated risk management framework
is a description of streams of accountability and reporting that will support
the risk management process[4]:

> Risk management frameworks are a description of an organizational
> specific set of functional activities and associated definitions that specify
> the processes that will be used to manage risks.

The key idea behind integrated risk management is that the overall risks
of a company are managed in aggregate, rather than independently. The idea
dates back to the early 20th century writings of Henry Fayol. However, his
work has little penetrated the risk management profession, thus the notion
has been erroneously hailed as both a recent and a new paradigm[5]:

> The new paradigm is to broadly view risk management as an integrated,
> strategic, and enterprise-wide activity that involves employees at all lev-
> els of the organization. Risk management is coordinated with senior-
> level oversight, but everyone in the firm views risk management as an
> integral and ongoing part of their job.

The central objective is to provide senior management with risk information to assist in the optimization of the portfolio of business opportunities.
Correctly implemented, such a monocle on risk aligns corporate risk appetite

SOME OBJECTIVES OF INTEGRATED RISK MANAGEMENT

An integrated risk management framework should offer a practical guide to ensure relevant risk-related information is regularly collected and communicated in a timely manner throughout the organization. Core objectives include:

- Provide the setting against which control activities to assess, manage, and mitigate risk are designed and implemented.
- Strengthen risk management practices within risk-taking business units.
- Promote a company-wide view of risk grounded in the principles of responsible risk taking.
- Foster corporate citizenship and transparency by placing an emphasis on consultation and inter-business unit communication.

with stated performance objectives and embeds risk management deep into corporate culture. Risk-return decisions taken in underlying business units can be made with reference to the framework. This, in turn, can provide greater transparency about the alignment of risk and strategic objectives to the chief executive officer (CEO) and board of directors.

The Hidden Dangers of Risk Management Silos

The traditional approach to risk management is by its very nature tactical rather than strategic. This is because decisions about managing risk are usually taken at the business unit level without consideration of the risk management activities carried out in other areas of the company. For example, the treasury of a British airline might decide to tactically hedge the price of jet fuel oil.[6] By using forward contracts, the airline locks in the Sterling cost of its jet fuel purchase. The decision to hedge or not may be made without any consideration of other hedging or insuring activities carried out in other areas of the company. This may be so even when the risks across units are significantly correlated[7]:

> Historically, and still in many organizations, the paradigm was for risk to be managed in "silos." Silos typically exist when disparate areas of the business are managed as narrowly focused and fragmented activities. For example, silos can exist for insurance, foreign exchange risk, operational risk, credit risk, and commodity risk.

The consequences of failure to monitor risks at the aggregated level are illustrated by the case of catalytic converters at the Ford Motor Company. Risk writer Jack Shaw describes what happened[8]:

> In the early '90s, two different areas at Ford recognized that a significant price increase in some or all of these rare elements [platinum, palladium, and rhodium required for the production of catalytic converters] could have a devastating impact on Ford's profitability. Each set out to do something to reduce this risk. The purchasing department took a commodities hedging approach to solving the problem. They reduced the risk by entering into a series of long-term contracts to purchase these rare metals at prices locked in to the acceptable prices in the market at the time. The research and development department took a research-based approach to solving the problem. They determined to develop new catalytic converters that required only a tiny fraction of the rare and expensive metals in question. ... Five years later, Ford had new catalytic converters that no longer required the metals they had committed to purchasing. Unfortunately, at that time, an oversupply of these metals in the marketplace had brought about a price drop. As a result, Ford had to take a write-off of almost a billion dollars on their rare metals purchase contracts.

The fact that Ford's various risk overseers did not share risk information lay at the heart of their troubles. A firm-wide integrated risk system would have allowed Ford to better manage this particular risk by aiding communication between the various departments and encouraging the company to develop a strategic solution to the problem. This may have involved using a 1- or 2-year horizon for financial hedges, while over a longer horizon, making operational adjustments such as alternative sourcing, utilization of different plant locations, pricing, and the outcome of the research and development department.[9] During early 2000, senior management at Ford decided to set up a global risk management group. It was headed by a new position, director of global risk management.[10]

The ability to locate and exploit natural hedges can yield a vital competitive advantage. Yet, the silo approach to risk management continues to persist; risk management departments continue to ignore natural hedges[11] that occur during the normal operation of business. Managing risk in silos is likely to prove a poor long-term business strategy. Occasionally institutions can make an avoidable mistake, such as suffering losses from inappropriate hedging, and recover. However, second and third mistakes can bring into question a company's business strategy and risk management practices. This is especially the case in today's hyper-competitive environment characterized by heightened risk of shareholder litigation, accounting scandals, and increased media and regulator scrutiny. As the obsessive megalomaniac genius, Auric Goldfinger, famously stated[12]:

Once is happenstance. Twice is coincidence. Three times is enemy action.

In cases perceived by the market as enemy action (i.e., poor management), a company can expect to experience a savage assault on its senior management team, precipitous decline in share price, and possibly investigation by regulators and legislators. The share price may well remain depressed until the institution reestablishes its credibility or is taken over.

The Need for Better Risk Management

Integrated risk management aids the process of identifying and estimating the financial impact and volatility of a company's unique portfolio of risks. Thus, in addition to helping companies identify, measure, and monitor risk, it can help prevent losses by introducing solutions that boost shareholder value. As the Federal Deposit Insurance Corporation (FDIC) noted in its discussion of the U.S. savings and loans crisis of the 1980s, superior risk management characterized the survivors[13]:

> Furthermore ... although banks that failed had generally assumed greater risk before their failure, many other banks with similar risk profiles did not fail. In the case of these surviving banks, the effects of risk taking, including risk taking stimulated by under priced deposit insurance, were apparently offset by other factors, including superior risk-management skills. The absence of these offsetting factors should therefore be considered more important causes of bank failures.

The need to manage risk more coherently through integrated risk management is therefore more critical than ever. An integrated risk management framework is the base upon which a more strategic and corporate-wide approach to risk management can be founded. It serves as the unifying structure that overrides and directs risk management practice. It concords with the general notion that[14]:

> Good risk management is now increasingly seen as an integral part of "business as usual" rather than a set of discrete activities, carried out with the regulator in mind.

It also explicitly defines how the management of risks is to be handled by:

1. Providing a logical and systematic method of identifying, analyzing, evaluating, treating, monitoring, and communicating risks associated with any business activity, function or process.
2. Setting the backdrop against which risks are managed in business units, in terms of how they will be identified, analyzed, controlled, monitored, and reviewed.

3. Providing information on roles, responsibilities, processes, procedures, and standards (in the form of policy and procedural guidance) to guide managers and other employee's compliance.

The Challenge

The challenge, therefore, is for companies to approach risk management in a more integrated and systematic way that includes greater emphasis on consultation between business units and communication with shareholders, the board, and regulators. A systematic and integrated but adaptable approach to risk management requires an organization to build capacity to address risk explicitly. It will only emerge if risk accountability and awareness is driven deep into the cultural roots of an organization.

Until very recently senior executives have been somewhat diffident in pushing risk management up the corporate agenda[15]:

> Today's global business leaders know that the chances and potential cost of a risk management failure or "near miss" in their organisation are too high, and they are devoting much more time to formal risk management now than they did three years ago, and are assessing a wider range of threats. Nevertheless, competition with other business priorities presents a real obstacle to embedding risk management throughout organisations. The fact that boards are only slowly becoming conscious of the connection between good risk management, better financial performance and stronger corporate reputation suggests that they need to focus more closely on the wider benefits of fully integrating risk management into corporate decision-making, and on the tools available to facilitate this process. Until they begin to do so, risk management is likely to continue to be seen by senior management as a constraint on their business rather than as a source of competitiveness.

A survey carried out by Lloyd's of London just before the financial crisis of 2008 reported that, despite spending more time on risk management, board directors at global business were still failing to identify and adequately manage emerging risks.[16] Around the same time, the United Kingdom's Financial Services Authority, in a review of insurer's risk management processes, noted that[17]:

> ... weaknesses remained, including:
>
> 1. a failure to capture risk management information across all of its risks;
> 2. risk information being partly held centrally and partly locally;

3. a lack of integration of the firm's "bottom up" and "top down" processes for identification; and
4. the risk management function was not involved in all of the firm's major projects.

> … This inadequacy meant that much risk management effort was wasted, either through the lack of appropriate information or in the development of "workarounds" to overcome the lack of central capability.

The lack of commitment to effective risk management may be partly because in an environment where every expenditure has to have a revenue payback, it can often prove difficult, prior to a catastrophe, to convince senior management that a cultural change in risk management is really necessary. One hopes the cumulative effect of documented risk management failures such as the case of catalytic converters and the Ford Motor Company, failure of Lehman Brothers, and the record number of U.S. bank failures that occurred between 2008 to 2011, will add urgency to the drive for more efficient risk management. However, it is more likely that changes in the regulatory environment and or requirements of rating agencies[18] will have a more speedy impact. Fitch, Moody's, and Standard and Poor's[19] have all developed processes for assessing how well executives understand and manage risk, the strength of the risk management function and extent of senior management oversight of operational and financial risks.

An integrated risk management framework can play a positive role in buffering earnings from avoidable losses and thereby preserve value by avoiding damage to a company's reputation. Hence, whether it is avoiding the real cost of breakdown or failure of a business process or redirecting the opportunity cost of fixing problems to more value-added activity, a comprehensive and vigorously applied integrated risk management framework supports better decision making by contributing greater insight into strategic and tactical risks.

In 1999, Hydro One, the largest electricity delivery company in Ontario, Canada, appointed an individual by the name of John Fraser into the role of chief risk officer (CRO). Fraser a sharp-eyed chartered accountant was given a mandate of six months to show the position and enterprise risk management [ERM] could deliver. ERM was a success[20]:

> … one of the best examples of a quantifiable benefit of ERM at Hydro One is lowering the cost of debt. To illustrate—In 2000, Hydro One issued $1 billion of debt, the first issue as a new company after the demerger from Ontario Hydro. According to conversations with senior ratings analysts at Moody's, ERM was a significant factor (and continues to be a significant factor) in the ratings process for Hydro One, although it is difficult to quantify the benefit. It is important to note that the firm received a higher rating than initially anticipated (AA– from S&P and A+ from Moody's) on this issue and the issue was heavily oversubscribed by approximately 50 percent. To quantify the potential yield savings, consider that since 2000, the long-term mean yield spread between AA and

A has averaged approximately 20 basis points. On $1 billion of debt, this results in an annual savings in interest costs of approximately $2 million (i.e., $1 billion × 0.0020). While it is impossible to estimate the interest savings attributable directly to ERM, clearly ERM was a beneficial factor as noted by the analysts.

The Three Essential Elements of Successful Risk Integration

The question that naturally arises is, "what are the essential elements of integrated risk management?" Guidance on this issue is rapidly expanding. Canadian scholars Shortreed, Craig, and McColl[21] in a comprehensive review identified over 80 distinct generic integrated risk management frameworks and guidelines including:

1. In the United Kingdom, the Institute of Risk Management, Association of Insurance and Risk Managers, and The National Forum for Risk Management in the Public Sector published a joint risk management framework.[22]
2. The Canadian Standards Association's (1997) generic risk management framework.
3. The Japanese Standards Association (2001) risk management guidelines.
4. In the United States, the Committee of Sponsoring Organizations of the Treadway Commission framework for enterprise-wide risk management.[23]

This bewildering and ever growing assortment is symptomatic of the fact that one size does not appear to fit all. Differences in the way in which risk is managed through organizational structures[24] are inevitable because of the wide variety of business processes throughout modern corporations.[25] Thus, while it is true that the development of a risk management framework is heavily dependent on the nature of the business, the guiding principle is that the specific framework implemented be comprehensive enough to capture all significant risks a company is exposed to and has flexibility to accommodate any change in business activities. At the same time, since in any business there are multiple levels of risk-taking activity, the framework, while providing a realistic representation of the risk management behavior, needs to be transparent enough to be communicated and understood both within and outside of the organization. Given the complexity involved, the chief architect needs to be a dedicated risk professional—the chief risk officer.

It would be difficult for any individual to have detailed knowledge of all of the existing frameworks, much more so for a business unit head, senior executive, or board member. Indeed, many professionals involved in the business of risk management, such as auditors, actuaries, and risk managers are ignorant of frameworks proposed by their fellow professionals. What is critical to grasp, is not so much the detail of any specific framework,[26] but what these frameworks have in common[27]:

1. Risk identification, assessment, monitoring, and modeling.
2. Risk management.
3. Strategic risk setting.

These three elements provide the commonality between frameworks; they lie at the heart of successful integrated risk management.

Tying the three elements together in a coherent fashion requires the development of a set of clearly defined risk management policies and procedures covering areas such as risk tolerance, identification, acceptance, measurement, monitoring, reporting, and control. In general, the resultant risk management policy document will begin with a statement of the company's attitude toward risk.

Alongside details of a firm's risk tolerance will be the risk management objectives and clearly defined risk limits. This process will involve setting out the ground rules for what areas of risk should be transferred outside of the organization rather than managed internally; a detailed description of the criteria used for measuring and managing relevant risks (market, credit, liquidity, operational, counterparty, etc.), in addition to details of reporting formats and procedures.[28] The risk management policy document should obtain board approval before being implemented. It may also be supplemented with various risk management committees, which may oversee specific risks facing an institution.[29]

The control structure arising from a risk management framework is a vital element in providing assurance to boards and shareholders. Since in practice, risks are the operational responsibility of the business unit from which they are generated, it is critical to make sure alert signals are effectively measured, monitored, assessed, and treated uniformly across the entire organization. Confidence that the process is functioning adequately requires management to establish ongoing monitoring of performance. Reviewing the effectiveness of risk management oversight is also important; at the very least, the entire framework itself should also be subject to annual review covering areas such as whether management is satisfactorily enforcing existing controls, adequacy of the current process for supervising risk control process, and concerns (if any) over inadequate separation of duties. Such reviews provide confidence that the objectives of the risk management

**WHAT STEPS SHOULD I TAKE TO IMPLEMENT
A RISK MANAGEMENT FRAMEWORK?**

Step 1: Establish and document clear roles, responsibilities, and accountabilities for the members of the risk team including the chief risk officer.

Step 2: Develop guidelines for managing risk, ideally through a risk policy document. Set up and scope various risk management oversight committees. Identify and evaluate investment and other risks in products and portfolios.

Step 3: Review and set suitable limits on each product/portfolio.

Step 4: Ongoing monitoring of the framework with frequent reviews of its applicability, achievements, and weaknesses.

framework are being achieved and that control activities are operating effectively. The results should also be presented to the board.

Fundamentally, the decision to build an integrated risk management framework comes down to the fact that it can create and protect real competitive advantage. Of course, going from desires to actions requires much more than documentation and corporate value statements, it also requires a tangible and visible commitment from senior management, executive-level input, and oversight. Changing the risk management culture of a corporation requires inspired leadership and gutsy decisions because it requires genuine commitment from within business units to be fully effective.

Yet, a monocle on risk can provide an important foundation upon which to advance a more systematic approach to risk and lay the foundation for a well-functioning risk management culture. One in which the decision processes surrounding new products and services are documented, open, and transparent; where a standard set of terminology is used to describe risk issues, thus contributing to more effective communication.

For Further Thought

This chapter focused on the underlying ideas surrounding integrated risk management; the primary objective of which is to measure and manage risk across a range of diverse business activities. The chapter did not discuss in detail any specific frameworks (as we saw these are numerous). However, all frameworks have a common core that boards, senior executives, and

INTEGRATED RISK MANAGEMENT

Key Point 1: Risk is an inherent part of doing business. Its adequate management demands every decision maker evaluate their risk-bearing activities by reference to risk management principles.

Key Point 2: Traditional risk management is tactical rather than strategic, with decisions about managing risk taken at the business unit level. Little consideration is given to the risk management activities in other areas even if risks across business units are highly correlated.

Key Point 3: Key to infusing a *risk scent* into the cultural fabric of an organization is the notion of an integrated risk management framework.

Key Point 4: Integrated risk management is an alternative to traditional risk management in which overall risks are managed in aggregate, rather than independently.

Key Point 5: An integrated risk management framework is a description of an organizational specific set of functional activities that specify the processes that will be used to manage risks.

Key Point 6: Integrated risk management has a number of objectives including:

1. To provide the setting against which control activities to assess, manage, and mitigate risk are designed and implemented.
2. To strengthen risk management practices within risk-taking business units.
3. To promote a company-wide view of risk grounded in the principles of responsible risk taking.
4. To foster corporate citizenship and transparency by placing considerable emphasis on consultation and interbusiness unit communication.
5. To offer a practical guide to assist in day-to-day decision making.
6. To ensure relevant information is regularly collected and communicated throughout the organization.
7. And to optimize the firm-specific portfolio of business opportunities to achieve stated performance objectives.

Key Point 7: It defines how the management of risks is to be handled by:

1. Providing a method for identifying, analyzing, evaluating, treating, monitoring, and communicating risks.

2. Defining how risk will be identified, analyzed, controlled, monitored, and reviewed.
3. Providing information on roles, responsibilities, processes, procedures, and standards.

Key Point 8: The integrated risk management framework should also be subject to periodic review. The results should be presented to the board.

Key Point 9: In an environment where every expenditure has to have a revenue payback, it can be difficult, prior to a catastrophe, to convince senior management that the cost associated with forcing through a cultural change in risk management is really necessary.

management will need to consider in developing their own firm-specific framework. This will include four basic issues; specifically how to:

1. Increase a company's understanding of its overall portfolio of risks.
2. Provide management with a better understanding of those risks.
3. Identify and assess the major risks and create consistent, firm-wide solutions for dealing with them.
4. Manage risk to reduce the impact of losses on the balance sheet or minimize variability in earnings.

We set out below a series of questions which senior management may wish to consider when reviewing the current state and effectiveness of their own integrated risk management framework:

- Are the key high-level risks for the business unit known?
- What is the available evidence that the business unit heads are engaged and committed to the corporate risk profile?
- What resources have been allocated to risk management?
- Is there sufficient capacity to manage risk within the organization? How do you know?
- Is risk an integral part of the decision-making processes?
- Are employees aware of risk management limits and practices?
- Are systematic risk management processes already being applied?
- Do current employees have the necessary knowledge, skills, and tools to manage risks given their areas of responsibility?
- Is there a designated risk champion for overall corporate risk?

- Are there designated risk managers or processes within each business unit?
- Has the organization adopted a common process for risk management?
- Is standardized risk management terminology communicated, understood, and applied throughout all relevant organizational processes?
- Are automated systems and processes in place to monitor risks?
- Are processes in place to support periodic communication with all stakeholders on risk tolerances and actual levels of risk?
- How are risk management practices and procedures integrated into preexisting governance and decision-making structures?
- Have audits of the risk management process been conducted, if so have actions been taken on their findings?

In addition to the above questions, we also list below the 10 questions for senior management the Financial Services Authority prior to the 2008 financial crisis identified as critical when reviewing the effectiveness of risk management practices[30]:

1. How can the board and senior management provide more effective and informed oversight of your firm's risks?
2. Are risk considerations given appropriate profile in your firm's business and strategic planning processes?
3. What should your firm be doing to realize the benefits of further integration of risk, capital, and business management activities?
4. How can your firm improve the knowledge and understanding of your board and senior management to raise the quality of discussion and challenge on more complex matters?
5. Are your firm's risk appetite statements and risk policies sufficiently comprehensive and well understood and workable?
6. Does your firm have a clear view of how it wants to develop its risk management practices?
7. Are there enough opportunities for independent and informed challenge to risk management processes and outcomes?
8. Is there enough objectivity in your risk identification and assessment processes?
9. Does your firm's management information provide sufficient and timely material on risk issues and does it prompt appropriate action?
10. Is there enough clarity on how responsibilities for risk management activities are allocated in your firm?

Additional Resources

Additional reading around the role a monocle on risk can play in protecting value can be found in Shortreed, Craig, and McColl (2003) and Aabo, Fraser, and Simkins (2005). The Federal Deposit Insurance Corporation (2000), Lloyd's of London and Economist Intelligence Unit (2005), Shaw (2005), and the Financial Services Authority (2006) all highlight the inherent flaws in traditional risk management. Specific national frameworks can be found in Canadian Standards Association (1997) and Japanese Standards Association (2001). A practical example of the highly praised risk management framework used by the BMO Financial Group can be found at What's Next? 187th Annual Report 2004 (http://www2.bmo.com/ar2004/downloads/bmo_ar04.pdf). Read the sections on risk management; it is very impressive, little wonder the BMO Financial Group was judged to be Canada's Best Corporate Citizen of the Year for 2005 by *Corporate Knights—The Canadian Magazine for Responsible Business*. Finally, as an illustration of a risk management best practice charter see the Irish Association of Corporate Treasurers Charter of Best Practice in Treasury Management available at their Web site (http://www.treasurers.ie/).

Aabo, T., Fraser, J, and Simkins, B.J. (2005). The Rise and Evolution of the Chief Risk Officer: Enterprise Risk Management at Hydro One. *Journal of Applied Corporate Finance* 17(3) (June).

Canadian Standards Association. (1997). *Risk Management: Guideline for Decision-Makers* (CAN/CSA-Q850-97). Rexdale, Ontario: Canadian Standards Association.

Economist Intelligence Unit. (2005). *The Evolving Role of the CRO: A New Report.* Economist Intelligence Unit. www.eiu.com/cro.

Federal Deposit Insurance Corporation (2000). *An Examination of the Banking Crises of the 1980s and Early 1990s,* Vol. 1. Federal Deposit Insurance Corporation: Washington, DC. http://www.fdic.gov/bank/historical/history/vol1.html.

Financial Services Authority. (2006). *Insurance Sector Briefing: Risk Management in Insurers.* Financial Services Authority: London (November).

Japanese Standards Association. (2001). *JIS Q 2001: Guidelines for Development and Implementation of Risk Management System.* Japan: Japanese Standards Association.

Lloyd's of London and Economist Intelligence Unit. (2005). *Taking Risk on Board: How Global Business Leaders View Risk.* London: Lloyd's of London.

Montier, J. (2010). *The Little Book of Behavioral Investing.* Hoboken, NJ: John Wiley & Sons, Inc.

Shaw, J. (2005). Managing All of Your Enterprise's Risks. *Risk Management Magazine* (September). Risk and Insurance Management Society.

Shortreed, J.H., Craig, L., and McColl, S. (2003). *Benchmark Framework for Risk Management* Final Report. Toronto, Canada: Network for Environmental Risk Assessment and Management.

Endnotes

1. Legendary value investor, Jeremy Grantham, speaking of the crash of 1974. See Montier (2010) for further details.
2. First published in 1945. See Dahl, R., "Madame Rosette," In *Collected Short Stories*, Jameson, E. (ed), (London: Penguin, 1993).
3. *Integrated risk management* is one of the most widely used out of a group of synonyms that describes a broad and comprehensive view of managing risk across the firm. Other terms used to describe this idea are *enterprise-wide risk management, enterprise risk management* and *firm-wide risk management*.
4. Shortreed, Hicks, and Craig (2003).
5. Aabo, Fraser, and Simkins (2005).
6. Jet Fuel Spot Prices are quoted in U.S. dollars.
7. See Aabo, Fraser, and Simkins (2005).
8. See Shaw (2005).
9. In fact, it is common practice among large corporations to use a combination of operational hedges (production, marketing, research, etc.,) to manage long-term exposure and financial hedges such as forwards, futures, and options to manage short-term exposure.
10. An individual by the name of Freeman Wood.
11. A natural hedge is the reduction in risk that can arise from a company's normal operating procedures. A British airline with significant sales in the U.S. holds a natural hedge to its currency risk if it also generates U.S. dollar denominated expenses.
12. In the suspense novel, *Goldfinger*, by Ian L. Fleming.
13. See the Federal Deposit Insurance Corporation (2000). Between 1980 and 1994, more than 1,600 banks insured by the Federal Deposit Insurance Corporation were closed or received Federal Deposit Insurance Corporation financial assistance. This was the largest number in any period of the 20th century since the creation of federal deposit insurance in the 1930s.
14. See Financial Services Authority (2006).
15. See Lloyd's of London and Economist Intelligence Unit (2005).
16. See Lloyd's of London (2005).
17. See Financial Services Authority (2006).
18. Also note that a large number of firms traded on various stock markets use the debt markets to raise capital. The cost of capital for a particular firm is a function of the perceived level of credit risk. Rating agencies assign credit ratings to capture this perceived risk. Since the agency ratings have an impact on the cost of capital, companies place a high degree of importance on maintaining or improving their credit ratings.
19. See FitchRatings (http://www.fitchratings.com/), Moody's (http://www.moodys.com/), and Standard & Poor's (http://www.standardandpoors.com).
20. See Aabo, Fraser, and Simkins (2005).
21. See Shortreed, Craig, and McColl (2003).
22. See Airmic (www.airmic.com) for further details.
23. See Committee of Sponsoring Organizations of the Treadway Commission (www.coso.org/) for further details.

24. Including to some extent risk measurement and monitoring.
25. For this reason, the author is generally opposed to the prescription of any specific risk management framework. The preference is for a more principles-based approach in which senior management choose the most cost effective or innovative approach to managing their portfolio of risks.
26. Because, in practice, organizations may cherry-pick from the various frameworks for which they are familiar or create their own unique framework from scratch.
27. Which involves setting the risk appetite of the organization and other performance objectives.
28. Reporting formats and procedures will differ depending upon whether the information is required for senior management, board of directors, auditors, or regulators. However, for each of these groups, risk management policy should at the very least define the layout, timing, and responsibility for production of risk reports, alongside the level of detail required.
29. In addition to a risk group, internal audit, and so on, a company may institute a risk management committee to supervise overall risk management at an institution. They may also set up subcommittees to oversee particularly complex operations such as those associated with derivative instruments and structured products. Such committees serve to entrench regular reporting cycles for the evaluation of key risks.
30. See Financial Services Authority (2006).

8

Benefit from the Fable of Spreadsheet City

You may have already come across my *fable of Spreadsheet City*, in which a small financial institution in the rural Midwest of the United States of America employs an intern by the name of Derek to develop and maintain a risk management system. Derek, fresh out of college and recalling a similar project he completed for his course on quantitative modeling, chooses to develop the system in a spreadsheet package, with additional functionality created through the spreadsheet's macro language. Initially a 1-day a week task, the system soon grows to such an extent that it becomes the primary focus of a team of highly paid programmers (who are brought in, at great expense, from the "Big City" several hundred miles away). The programmers enhance the macro language by coding pricing and numerical functionality in the programming language known as C++.[1] As time goes by, the head of human resources marvels at the contented team spirit of the programmers. Indeed, she notices that, despite being from the Big City and therefore somewhat adverse to the ways of country folk, they appear to be happier with each passing day.[2] She decides to retrain as a programmer and eventually joins the team.

Programmers generally require direction from quantitative and risk management staff, who, in turn, require support staff and project managers and so on. Slowly, a community of individuals dedicated to supporting the support staff who are supporting the people who are developing and maintaining the spreadsheet system grows up. With these individuals and their attendant families comes demand for community services—new roads, housing, schools, hospitals, and churches. Before long, the rural backwater is transformed into a small town. A mayor is elected (who in my favorite version of the story is the intern). In recognition of his good fortune and the booming local economy, the mayor declares the town be henceforth known as *Spreadsheet City*.

All goes well for the newly incorporated town, that is, until the day of *critical shut down*. You see, over time the spreadsheet had become so large and complex that no single individual could possibly understand it (the mayor had long since given up, devoting his time instead to city matters). The large and ever growing body of programmers were simply implementing specifications, project managers concerned with keeping things moving, and the multiplicity of end users focused on their specific tasks. No single individual or group was in charge of design, development, or implementation. The scope, reach, and criticality of the spreadsheet was unknown. Eventually, creaking under its own complexity the spreadsheet crashed.

Teams of programmers, management consultants, auditors, and project managers were brought in to address the shut down. Alas, to no avail. The spreadsheet system had been developed so quickly and easily without a consistent and thorough design methodology or test and documentation that it was impossible to untangle. Numerous spreadsheet amendments compounded the problem, as the majority had been made in an ad hoc manner without any documentation. With such a huge, convoluted, complex tangle of interlinking spreadsheets, ad hoc coded modules written in a surprisingly large variety of curiously named programming languages (ABC, Befunge-97, BLISS, BLoop, C, C++, CHILL, Euphoria, Hope, LIFE, RUBY, SNOBOL and SPITBOL, UFO, and ZPL), it was impossible to determine what precisely caused the failure. With the spreadsheet dead, the programmers began to leave town followed in quick succession by management consultants, project managers, auditors, other support staff, and their families. Within a few short months, Spreadsheet City lay deserted, empty, except for the mayor and a handful of diehards who continue to this day to tinker with the spreadsheet in the vain hope of bringing it back to life.

My fable of Spreadsheet City serves as a cautionary tale to the businesses, large and small, who have traditionally relied on spreadsheet-based systems for risk management and other mission critical functions.[3] Indeed, in the not so distant past, it was not uncommon for financial institutions to hire a graduate student or two to build a risk monitoring system in a simple spreadsheet application. As the range of traded financial products expanded and more complex analysis became necessary so did the risk management spreadsheets. The difficulties began to creep in, as the spreadsheet became a victim of its own success. Designing and developing new numerical routines for risk measurement in a spreadsheet package is generally rather straightforward. The use of macros combined with multiple linked spreadsheets allowed risk analysts to build very complicated, often convoluted, models with minimal or no documentation. Spreadsheets were linked to other spreadsheets, which, in turn, were linked to other spreadsheets. Linked spreadsheets would proliferate like rabbits until a business line's risk managers were dependent on them to support even the most basic reporting requirements. Ferreting out numerical quirks and bugs in complex linked spreadsheets can be rather arcane. The process of implementing and validating even simple changes can end up taking days or weeks to complete because complex linked spreadsheets are not the optimum forum for maximizing risk management performance. As performance of such systems falters, so does the objective of providing timely and accurate risk management information. Indeed, while spreadsheet developed applications often perform a vital role, unlike most professionally written applications, they are generally not created with a diligent requirements specification, code review, and then submitted to a thorough testing process. As we shall see, relatively simple spreadsheet errors can have a material impact on the bottom line. Given the central role of spreadsheets in business life, it is somewhat surprising that so little has been written about their associated risks. This chapter attempts to redress the balance by exploring aspects of spreadsheet risk management.

Don't Be a Victim of Spreadsheet Hell

Computing systems developed by users on their own to support their functional activities as employees began in the late 1970s, initially spurred on by the introduction in 1973 of the Xerox Alto personal computer, and later rapidly gathering pace when IBM, in the late summer of 1981, began production of the IBM 5150. During the 1980s, development of applications by end users accelerated rapidly, especially among users of spreadsheet and database applications.[4] Growth was driven primarily by ease and speed of spreadsheet development, flexibility, user control of programming, and the perceived ability of user-developed systems to assist their user developers to carry out their job function more effectively.

Panko (1988) in one of the first textbooks to address the subject of *end-user computing* reported that a significant number of business managers used spreadsheets during the 1980s to enhance their decision making. Today, spreadsheet use by managers is universal. They can transform manually tedious and time-consuming everyday tasks into quick and easy electronic tasks. Managers and others automate virtually every aspect of their job function via spreadsheets. Their natural interactivity, easy license-free development, and the universality of environment in which they can be run, offers such great advantages over traditional software development that they are not going to go away anytime soon.

Informal or personal spreadsheet systems are often run alongside corporate computer systems without being subjected to the same degree of control, quality assurance, or formal software development methodologies of the latter. Such systems can quickly become legitimized as part of the corporate knowledge base as other employees learn of their existence and come to rely on them as inputs to their own job functions. Yet, even very complex user-developed spreadsheets, although powerful software applications in their own right, tend not to be supported by the same control environment as formally developed software solutions. The systematic testing and the structured development methodologies of software engineering are not widely applied during spreadsheet development by end users. In addition, few of those who develop spreadsheet applications have formal training in structured programming, testing, version control, or software development life cycles.[5] As Croll (2005) comments:

> With the exception of quantitative finance professionals working as software engineers, there is almost no spreadsheet software quality assurance or appreciation of the software development life cycle as it might relate to spreadsheets. Spreadsheets built using well engineered code libraries were inevitably tinkered with later by traders, sales people, analysts and other users in an uncontrolled fashion.

In practice, many spreadsheets are poorly designed, difficult to understand, contain errors, and are inflexible. This has led to increasingly vocal concerns over their use in the workplace. Kavanagh (1997) writing in *Computer Weekly*, a heavyweight British computing magazine and several times winner of the United Kingdom Periodical Publishers Association's Campaign of the Year Award, cautions:

> End users are putting their companies at risk by setting up spreadsheets without realising that this demands the discipline of traditional programming. ... Our findings are disturbing ... as 78% of models (i.e., spreadsheets) had no formal quality assurance to ensure they were built to specified requirements and were fit for the purpose.

As spreadsheet models become more elaborate, they may eventually become too complicated for a new user to understand or maintain. Croll (2005) makes the point that:

> There is evidence that in the financial markets, spreadsheets are operating at, close to, or even beyond the present technological limits of their size and/or complexity. There were several reports of the present 256 column limit unnecessarily limiting the number of instruments in a financial portfolio and constraining the level of detail in temporal models. There were reports of difficulties in spreadsheets over 50Mb, a size which is not at all uncommon. Note that spreadsheets >1Gb already exist.

Given this, it is not very surprising that users of such packages soon become the victims of *spreadsheet hell*—complex, slow, error-prone systems which are difficult to maintain, change, or update.

Why Spreadsheet Failure Costs Big Time!

For those trapped in spreadsheet hell, ensuring any change is accurate and returns correct values is a difficult affair. Since all of the calculation details

IS YOUR BUSINESS A VICTIM OF SPREADSHEET HELL?

Spreadsheets are inherently easy to navigate and manipulate. Those tidy rows and columns of neatly aligned data, simplicity with which both small and large tasks can be achieved, instant updating, and sheer numerical horse power create a soothing illusion of orderliness, accuracy, and integrity. It is not always easy for executives with oversight of a range of business units to tell if a particular department is suffering from spreadsheet hell. For one thing, entrenched managers and their coworkers may be reluctant to admit to a problem. Second, unless you are heavily involved in the day-to-day operations, it can be difficult to have a real feel for any operational difficulties. Warning signs, which are observable from the outside include:

1. Someone in the department or business unit, usually with the word *analyst* in their title, spending most of their time managing spreadsheets that no one else can understand. The individual is probably seen as the spreadsheet guru. Problems associated with overreliance on this one person tend to emerge when they are on vacation, sick, or leave. If nobody else in the team can understand the spreadsheet, it is too complex.
2. Managers or analysts, and other coworkers are afraid to make even simple changes to a spreadsheet.
3. Individuals are rekeying information from or to a spreadsheet.
4. There exist a large number of spreadsheets, which support complex calculations, valuations, and modeling. These spreadsheets tend to be characterized by the use of macros and multiple supporting spreadsheets where cells, values, and individual spreadsheets are linked. In many cases, as in the fable of Spreadsheet City, you may find that nobody really understands their interdependencies or the underlying methodology.
5. Relatively standard reports, generated by spreadsheet, cannot be easily produced at short notice. Waiting too long for analysis is a sure sign of a problem somewhere along the line.

in a spreadsheet are generally exposed, they can rather easily (intentionally or in error) be corrupted. And because there is no audit trail on changes, mistakes may not be easily detected. The truth is that end-user developed spreadsheets are not as reliable as "software engineered" products and systems. Furthermore, the consequence of spreadsheet errors can cost senior executives such as the chief executive officer (CEO) their job. Under the title *"Mis-recording a number is a material weakness, lose your CEO position,"* the European Spreadsheet Risks Interest Group[6] reported:

Shares of RedEnvelope Inc. tumbled more than 25 percent Tuesday after the on-line retailer of specialty gifts drastically reduced its fourth-quarter outlook and said its chief financial officer will resign in April. Stanford Group analyst Rebecca Jones Kujawa said in an interview. "... they were underestimating the cost of goods sold ... it is likely CFO Eric Wong is being pushed out because of this error, which could demonstrate a material weakness in controls over financial reporting, an issue that usually leads to a lengthy review of accounting practices." RedEnvelope spokeswoman Jordan Goldstein said the budgeting error was simply due to a number mis-recorded in one cell of a spreadsheet that then threw off the cost forecast ...

Ultimately, the CEO is responsible for risk management. Eric Wong appears to have paid the price for failing to ensure risk management discipline, through negligence or lack of imagination, or both. Risk may be part of business life, but that hardly excuses failures on the part of management to properly access, manage, and monitor it.

Unfortunately, the RedEnvelope fiasco appears not to be an isolated incident. Spreadsheet malfeasance and malpractice have played a role in a large number of high-profile news stories[7]:

1. *TransAlta Corporation*: June 3, 2003, TransAlta Corporation, a Canadian power generator, announced a $24 million charge to earnings after a "clerical error" in pasting values into a spreadsheet resulted in it purchasing more U.S. power transmission hedging contracts, and at a higher price, than it required. "It was literally a cut-and-paste error in an Excel spreadsheet that we did not detect when we did our final sorting and ranking bids prior to submission." (TransAlta chief executive, Steve Snyder)[8]

2. *HealthSouth*[9]: Two ex-HealthSouth executives admitted that they prepared a false spreadsheet for auditors that inflated HealthSouth's assets and made the company appear to be worth more that it was.[10]

3. *Fannie Mae*: Fannie Mae had to amend their financial reporting due to spreadsheet errors. The result was that the correction increased unrealized gains on securities ($1.3 billion), accumulated other comprehensive income ($1.1 billion), and total shareholder equity ($1.1 billion).[11]

4. *Fidelity's Magellan Fund*: There was a big flap recently over Fidelity's Magellan fund estimating in November that they would make a $4.32/share distribution at the end of year, and then not doing so ... During the estimating process, a tax accountant is required to transcribe the net realized gain or loss from the fund's financial records (which were correct at all times) to a separate spreadsheet, where additional calculations are performed. The error occurred when the accountant omitted the minus sign on a net capital loss of $1.3 billion and incorrectly treated it as a net capital gain on this separate

spreadsheet. This meant that the dividend estimate spreadsheet was off by $2.6 billion ...[12]

5. *National Australia Bank*[13]: A spreadsheet error caused the
National Australia Bank to write down the value of its US mortgage book by AUS$ 3 billion, affecting its share price.[14]

6. *Allied Irish Bank/Allfirst*[15]: Allfirst "Would not pay the US$10,000
for a direct data feed from Reuters to the risk control section."
Instead, they got Rusnak [a currency trader] to download his
Reuters feed into a spreadsheet. He then substituted links to
his private manipulated spreadsheet. The total losses hidden by
the fraud were almost US$700M.[16]

7. *Audit of The Colorado Student Loan Program*: A formula in the
spreadsheet picked-up the date, 12/02/98, and interpreted it
as a dollar amount, resulting in an error of $36,131. As a result,
the beginning balance of the Federal Fund was understated
by $36,131.[17]

Yet, unfortunately, even after disaster strikes, failure to learn and to dif-
fuse the lessons to the wider business community persists. The litany of
costly spreadsheet errors from Allied Irish Bank to RedEnvelope, Inc. and
the cluster of ongoing errors reported by the European Spreadsheet Risks
Interest Group is evidence of the continued failure of corporate business to
take spreadsheet risk seriously. There must be many other similar cases that
have not been brought to public attention due to fear of a negative impact
on reputation and ultimately the share price of the company involved. Be it
because of the shortcomings of the risk management profession, impotence
of academic commentators, or lack of sufficient attention of those responsible for corporate risk measurement, management, and mitigation. Whatever
the reason, spreadsheet risk has often passed under senior executive's radar
until it is too late. The reality is that spreadsheet risk is pervasive; it covers all
areas of modern corporations. As the philosopher, essayist, poet, and novel-
ist George Santayana (1925) observed:

> Progress, far from consisting in change, depends on retentiveness ...
> Those who cannot remember the past are condemned to repeat it.

Spreadsheet packages are useful for data entry, certain calculations, and
reporting; they were never designed to be a dedicated risk management,
financial reporting, or other mission critical system.

That few in the wider business community consider the impact on their
operations of spreadsheet risk cannot persist indefinitely. For one thing,
the regulatory environment surrounding corporate governance is rapidly
changing. Regulators and legislators are enforcing governance practices that
will inevitably encourage senior executives to consider spreadsheet risk. For
example, as a consequence of compliance with the U.S. Sarbanes-Oxley Act,
which requires corporations to have a well-controlled financial reporting

A LITMUS TEST FOR SPREADSHEET RISKS

Here is a 5-point litmus test to assess the strength of your management of spreadsheet risk.

1. Do you have a list of your mission critical spreadsheets?
2. Is there a person who has ownership for compiling and updating this list?
3. Is the methodology underlying each spreadsheet documented?
4. Does each spreadsheet contain end-user documentation?
5. Have you "dethroned" your spreadsheet guru with a broader-based group of competent individuals and introduced a process that leads to knowledge sharing rather than knowledge concentration?

Does your litmus paper remain blue once you immerse it in the test? Let's hope so. You may be able to overcome failing one, maybe even two points on the above list; more than that, start constructing your safety net now, because your spreadsheet risk management environment is likely to be highly acidic.

system, many executives will have created an inventory of critical financial reporting spreadsheets.[18]

How to Bring Spreadsheet Risk under Control

A natural first step to bring spreadsheet risk under control is to create an inventory of all business critical spreadsheets. In general, such spreadsheets will fall under the heading of operational, analytical, and financial. Operational spreadsheets are those used to facilitate, track, and monitor operational processes. Analytical spreadsheets are those used to support day-to-day management decision making, and financial spreadsheets are those used directly to determine the financial status of an organization.

At the very least, the inventory of critical spreadsheets should contain the spreadsheet type (operational, analytical, financial), filename, location, function, individual(s), and department responsible for its development or maintenance, end users (who may be a different department or individuals within the same department), frequency of use, the presence or absence of supporting documentation (methodology and end user) and the criticality of the calculations. The last point is particularly important; for example, in the case of a series of financial reporting spreadsheets, one will need to identify:

1. The interrelationship between the spreadsheets;
2. Their individual relationship with financial statement assertions;
3. Then finally, a critical score such as high, medium, or low for each spreadsheet.

Some of the risks associated with spreadsheets can be reduced and better managed by the use of regular audits, the objective being to test the internal consistency, logic, and accuracy of both formulas and output. In addition, such audits should also investigate the reasonableness of any assumptions used. Periodic auditing can provide reasonable assurance to senior management that critical spreadsheets do not contain material or logical errors. The frequency of such auditing should be specified in corporate spreadsheet policy.

A lack of corporate-wide guidelines is one reason end-user spreadsheet developers do not adhere to rigorous development and testing standards. Yet, such guidelines can offer real benefits, not just in terms of advice and guidance. They can also serve as a backdrop against which incentives to learn about spreadsheet best practice can be initiated and encourage an open, blame-free discussion about identified spreadsheet errors. To help better frame their guidelines, corporate risk managers need to be cognizant of a number of issues surrounding spreadsheet risk management. These include the characteristics of spreadsheet error and the nature of spreadsheet engineering.

Understanding the Nature of Spreadsheet Error

A survey by Panko and Halverson (1996) found evidence that around 90% of production level spreadsheets contained significant errors. Hall (1996) identifies eight further studies, which have reported error rates in spreadsheets in excess of 30%. The situation does not appear to have improved much over time as 8 years later, Lawrence and Lee (2004),[19] in an audit of financially significant spreadsheets, found material errors in every single one. In fact, human cognitive research informs us that errors in complex spreadsheets will never be totally eradicated. As Panko (2000) notes:

> Error research in a number of fields has shown that there is almost nothing that human beings can do a thousand times in a row without making an undetected error. In fact, it is common to have undetected errors in

about 0.2% to 0.5% of all simple actions, such as typing keystrokes, and in about 2% to 5% of all more complex human cognitive activities, such as writing lines of computer code. ... The Spreadsheet Research website has data from over a dozen experiments involving more than a thousand subjects ranging from rank novices to experienced professionals. In all of these experiments, at least 1% of all cells contained errors.

In addition, three cell-by-cell field audits of real world spreadsheets have found errors in the 1% to 3% range. ... Other field audits, which did not inspect all cells, found smaller error rates, but every field audit found material error rates in at least 10% of all spreadsheets audited.

Although the old adage that "it is impossible to make anything foolproof because fools are so ingenious" may have a humorous ring of truth, given the findings of human cognitive research, the required level of ingenuity may be somewhat lower in the case of complex linked spreadsheets. The issue, therefore, is not whether errors exist (because in all likelihood they do), but their materiality and how they can be reduced and better managed.

As spreadsheet error adversely impacts on the integrity and reliability of spreadsheet output, a natural starting point is to consider the types of error that can occur during spreadsheet development. Panko and Halverson (1996) and Rajalingham, Knight, and Chadwick (2000) among others have developed taxonomies of spreadsheet errors. Broadly speaking, these taxonomies identify at the highest level two fundamental sources of error; those that occur as a result of bugs in the underlying spreadsheet software and those generated by users. While the first type of error is generally outside the influence of the typical spreadsheet developer/end user, the second can be influenced by spreadsheet development practices.

Spreadsheet error taxonomies generally categorize user-generated errors into two broad groups—quantitative errors and qualitative errors. Quantitative errors are those errors in which the spreadsheet gives incorrect results. They include errors due to omissions, simple mistakes such as typing in a wrong number or pointing to the wrong cell, accidental alterations, and logical errors such as entering the wrong formula. Qualitative errors are those errors that do not produce incorrect results immediately, but may do so at a later stage. They primarily arise due to flaws in the design and layout of a spreadsheet.

The Principles of Spreadsheet Engineering

The popular computing dream of *programming without programmers* was the primary focus of fourth-generation computer languages. They did not truly achieve the dream (being limited to report generation from databases), however, the idea of programming without programmers proved extremely popular with spreadsheet users. When the first spreadsheets began to appear, 30 years or so ago, they were very hard to use as they required very specialized programming skills. Over time, however, they have become much more *user-friendly*, so much so, that today almost anybody can use a spreadsheet to design an application even if they have absolutely no knowledge of computer programming. Indeed, end users create and distribute their own spreadsheet applications in increasingly large numbers.

Unfortunately, as we have seen, programming without programmers has often resulted in programming without discipline. One idea to counter the growth of undisciplined spreadsheet applications, particularly in the corporate sector, involves "engineering" better spreadsheets using some of the tools and approaches of software engineering. To many, especially computer scientists, this is an inherently appealing approach because it considers spreadsheets for what they really are—complex pieces of software, which must be well designed if they are to function optimally. As Rajalingham, Chadwick, Knight, and Edwards (2000) note:

> Contrary to the traditional view that a spreadsheet is merely a flexible electronic worksheet, it should be viewed as a computer program. A close examination of spreadsheet structure would reveal that a spreadsheet is fundamentally similar to a computer program.

The "engineering" of a complex computer program typically consists of a requirements elicitation in which the objectives or purpose of the software are defined. This is followed by specification and design of the required functionality. It is only after this stage is complete with a full specification that the coding of the software begins. As the program is developed, it is subject to extensive testing to demonstrate it is fit for purpose. Documentation of the underlying program methodology and production of end-user documentation are also a vital part of the "engineering" process. Over time maintenance and updating of the program occurs as the needs of the end user changes. Since spreadsheet development mirrors much of these activities, the term *spreadsheet engineering* has been coined to describe the application of software engineering ideas to spreadsheet design and development. As Grossman (2002) explains:

> The application of software engineering principles to spreadsheets— call this "spreadsheet engineering"—has the potential to increase the productivity of spreadsheet programmers, decrease the frequency and severity of spreadsheet errors, enhance spreadsheet maintainability over time, and actually be implemented by spreadsheet users.

The formal discipline intrinsic in spreadsheet engineering encompasses all aspects of spreadsheet creation and embodies the idea that well functioning error-free spreadsheets do not happen spontaneously but rather that they must be consciously engineered. Grossman (2002) outlines eight guiding principles:

Principle 1: Best practices can have a large impact.

Principle 2: Life-cycle planning is important.

Principle 3: *A priori* requirements specification is beneficial.

Principle 4: Predicting future use is important.

Principle 5: Design matters.

Principle 6: Best practices are situation dependent.

Principle 7: Programming should be a social and not an individual activity.

Principle 8: Deployment of best practices is difficult and consumes resources.

Applying software engineering to complex spreadsheet development injects discipline into both design and development. Such rigor ensures that new users and developers can rapidly understand, maintain, and easily update an existing complex spreadsheet application.

The Potential of Compilable Spreadsheets

The ability to compile spreadsheets would allow complex spreadsheet applications to be developed along similar lines and with the same rigor as other more traditional software applications. The spreadsheet developer codes the application specification directly into a computer language analogous to C++ or FORTRAN. Once complied, the resulting spreadsheet contains the necessary functionality and can be used as any other spreadsheet. Computer languages, which compile directly into a spreadsheet application, also offer up the possibility of inverting the process through a decompiler. This would give the spreadsheet developer the ability to decompile existing spreadsheets into their underlying code. The code could then be subjected to the discipline of software engineering, with weaknesses being remedied and enhancements efficiently applied.[20]

There has been some academic and commercial activity in the area of compilable spreadsheets;[21] however, their use requires rather specialized programming knowledge. While this may be beneficial from a quality engineering perspective, ironically, it will hinder widespread acceptance because many of the existing end users lack modern programming skills or even the desire to learn them. Since the current popularity of spreadsheets lies in their ability to successfully shield users from the low level details of traditional programming, the role of compilable spreadsheets will likely always be somewhat limited. Nevertheless, where information technology departments are formally involved in the development of critical and complex spreadsheets, the use of a compilable spreadsheet language may confer significant design, testing, and maintainability advantages. In general, however, compilable spreadsheet languages offer at best a partial, but potentially valuable, solution to the development of complex spreadsheets.

Seven Rules for Superior Spreadsheet Design

A number of academic studies have investigated the relationship between the visual design of a spreadsheet and the accuracy of entered cell formula.[22] The evidence appears to suggest visual design matters. Poor visually designed spreadsheets tend to have more errors than those that take visual design issues into account. Researchers have attempted to identify a standard set of rules to be followed to produce well-designed spreadsheets.[23] Unfortunately, there is no one-size-fits-all solution, as the value of most rules appears to depend on the specific context.[24] However, it is possible to identify some general guidelines which are relatively easy to implement and widely applicable. Seven of these are outlined below.

Rule 1: Design before you build—With spreadsheet development, there is a tendency to begin building an application without a preliminary design. Yet, a critical lesson from the software engineering literature is that it is essential to have a clear idea of the purpose of the application to be developed. The problem should be clearly defined and the spreadsheet application thoroughly specified. The operations required to satisfy the purpose should be identified and their associated formulas documented. This includes an understanding of what user-entered data will be required and their expected ranges (checks to verify that input values satisfy these ranges should be a part of the automatic error checking of your application). Without a preliminary design, large complex spreadsheets are likely to be built in an ad hoc fashion with very many iterations. Good design will reduce the likelihood of error.[25]

Rule 2: Use Information Isolation—Different information types within a spreadsheet need to be isolated. For example, user-supplied parameter values should be separated from data, which, in turn, should be clearly separated from intermediate calculations and results. Where appropriate, formal cell protection should be applied within each

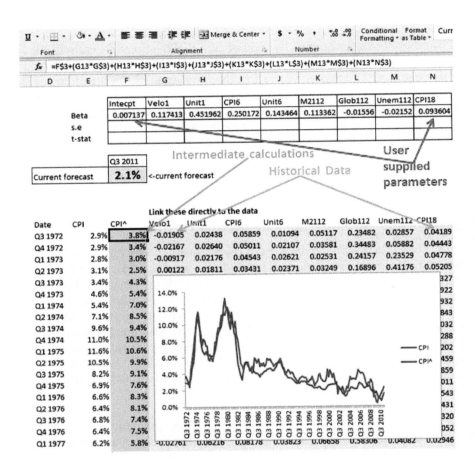

FIGURE 8.1
Formulated worksheet.

of these blocks. Visually, each block type should be clearly identifiable. It is common to use different colors, for example, user-supplied
parameters may be highlighted in yellow, intermediate calculations
in green, data blocks in red, and final results in gray (see Figure 8.1).

Rule 3: Use Named Cells and Ranges—The use of named cells and ranges
improves the readability of spreadsheets, especially in situations
where complicated formulas are used. For example, consider the following formula:

$$= (B16^2*D\$5^2 + ((1-B16)^2*D\$7^2) + 2*(D\$5*D\$7*B16*(1-B16)*D\$8))^0.5$$

Now, contrast the readability with the same formula, this time using
named cells:

$$= (\textbf{Weight\^{}2*Stdev\_Stock1\^{}2} + ((\textbf{1-Weight})\textbf{\^{}2*Stdev\_Stock2\^{}2}) +$$
$$2*(\textbf{Stdev\_Stock1*Stdev\_Stock2*Weight*(1-Weight)*Correlation}))\textbf{\^{}0.5}$$

While the above three rules are not groundbreaking, their consistent application has the potential to make spreadsheets more under-standable, manageable, and maintainable. The four spreadsheet rules of Raffensperger (2000)[26] provide additional design guidance:

Rule 4: Make your spreadsheet read from left to right and top to bottom.

Rule 5: Be concise with sheets, blocks, cells, formulas, and blank space.

Rule 6: Format for description, not decoration.

Rule 7: Show and describe your assumptions.

How to Minimize Risk through Formal Testing

The main objective of formal spreadsheet testing is to detect errors before the spreadsheet goes into widespread use. However, few end-user spreadsheet developers test their spreadsheets systematically, even fewer document the tests undertaken, even for mission critical spreadsheets. The most common approach to spreadsheet testing is undoubtedly execution testing in which the spreadsheet developer assembles a large number of test cases with known results and runs these test numbers through the spreadsheet application comparing the observed (spreadsheet) and known results. Where the observed output is identical to the expected output for all test cases, the spreadsheet is declared error-free. In its most basic form, execution testing involves finding errors by executing the spreadsheet application and seeing what happens. There are two basic kinds of execution testing; unit tests, in which the spreadsheet developer checks his or her own code to verify that it works correctly, and system tests, in which independent individual(s) check to see whether the spreadsheet application operates as expected.

Since a new spreadsheet application is generally developed cell-by-cell, for it to consistently produce the correct output, every input value and formula needs to be entered correctly. A complementary approach to execution testing involves the individual who developed the application looking at each cell to ensure the correctness of the entered formula or value. In mission critical spreadsheets with many tens of thousands of cells, individual code inspection is likely to be more efficient, both in terms of knowledge transfer about how the spreadsheet functions and also detecting errors.[27]

Team code inspection, long-recognized as a powerful way of reducing errors by the software engineering community, involves a team examining a spreadsheet application module-by-module. Typically, the work is carried

SPREADSHEET RISK MANAGEMENT

Key Point 1: Many businesses, large and small, have to rely on spreadsheet-based systems for mission critical functions. These spreadsheets are complex pieces of software.

Key Point 2: Many of these spreadsheets are developed by end users who have no formal training in computer programming.

Key Point 3: Unlike professionally engineered software applications, user-developed spreadsheets tend not to be created from a formal requirements specification or subjected to the rigor of software engineering testing.

Key Point 4: Underengineered spreadsheets are often poorly designed, difficult to understand, contain errors, and are inflexible. Spreadsheet errors can have a negative impact on reputation and ultimately the share price of the company involved. Human cognitive research suggests errors in complex spreadsheets will never be totally eradicated.

Key Point 5: The number and frequency of costly spreadsheet errors is evidence of the continued failure of corporate business to take spreadsheet risk seriously.

Key Point 6: Spreadsheet risk can be reduced and better managed by the use of regular audits to provide reasonable assurance to senior management that critical spreadsheets do not contain material or logical errors. The frequency of such auditing should be specified in corporate spreadsheet policy. At the minimum, such audits must test:

1. The internal consistency, logic, and accuracy of both formulas and output.
2. The reasonableness of any assumptions used.

Key Point 7: One reason end-user spreadsheet developers do not adhere to rigorous development and testing is that few corporations have specific policy or guidelines for end-user spreadsheet development. Such policy can offer important advice on planning, design practices, and provide incentives to learn about best practice.

Key Point 8: Software engineering principles can inject discipline into both design and development of spreadsheets.

out in two phases—an individual phase followed by a group meeting. Prior to the individual code inspection meeting, the developer(s) are required to explain the spreadsheet, how it relates to the overall problem to be solved, the underlying logic applied in the solution, identify areas of greatest risk of error (such as data capture areas, significant and complex formula, bottom line figures, and so on). The role of the individual and team inspection is to verify logic, formulas, check cell references, and so on. During the individual inspection, the team members independently examine the spreadsheet module-by-module looking for errors by investigating questions such as is the spreadsheet doing what the user intended? Are formulas logically and syntactically correct? Have any factors been omitted in the design? They then meet in the team phase to compare their results and also to search collectively for any remaining undetected errors.

For Further Thought

The ease with which malfeasance and malpractice can occur in conjunction with spreadsheet-based systems should be a powerful incentive for seeking out *off the shelf* automated systems. The gains in productivity, reduction in operational risk from the automation of manual processes alongside the ability of vendor systems to provide comprehensive pricing libraries, and quality checks of the imported data provide another powerful incentive to move critical operations away from spreadsheets developed by specific individuals. One must accept that sometimes there is no choice but to move beyond the familiar world of spreadsheets.[28]

Understanding how and where spreadsheets are used is key to controlling the associated operational risk. A corporate-wide spreadsheet risk management policy setting out best practice design may reduce the risk of spreadsheet errors and therefore enhance quality. Details of the required control processes, handling of critical spreadsheets, and the subsequent enforcement of corporate standards can also play a significant role in reducing the operational risk posed by end-user spreadsheet development.

1. Do you have policies and procedures in place to control spreadsheet use?
2. Do you have an inventory of business critical spreadsheets?
   i. Knowledge of their purpose and use?
   ii. Where they are kept?
   iii. Who has access and change rights?

3. Can you clearly distinguish between operational, analytical, and financial spreadsheets?
    i.   Are you able to rank them by their frequency of use?
    ii. Have you a sense of their inherent complexity (medium, low, high)?
    iii. Are you able to rank their criticality?
4. Have all of these spreadsheets been documented?
    - Complexity of the spreadsheet.
    - Details of the calculations.
    - Type of potential input.
    - Range of values and potential output values.
    - Underlying logic.
    - Size of the spreadsheet.
5. Have the spreadsheets been "software engineered?"
    i.   Formal elicitation of business requirements?
    ii.   What was the level of spreadsheet development experience of the developer(s)?
    iii. What was the nature of the testing carried out on the spreadsheet?
    iv.   Is the spreadsheet subject to formal version control?
    v.   Are you using any software that can log changes made to spreadsheets to supply an audit trail?
    vi.   Who created the end-user documentation?
    vii. How is the spreadsheet maintained, and who has responsibility for this?
6. For the critical spreadsheets identified, is there in place:
    - Clear directions on where they should be kept?
    - Who holds the keys to their design?
    - Knowledge of the end users (number and identity)?
7. These days it is important that your employees are familiar with spreadsheet design and development best practices.
    i.   Do you offer training to your employees in spreadsheet design and auditing?
    ii. If not, where do they get their best practice knowledge from?

Additional Resources

Santayana (1925), Panko (1988), and Taylor, Moynihan, and Wood-Harper (1998) provide historical context. News of costly or embarrassing spreadsheet errors is reported on the European Spreadsheet Risks Interest Group[29] Web site (www.eusprig.org). Also see Kavanagh (1997) and the report of the Office of Inspector General (2003). The fundamental nature of spreadsheet error is touched upon in Hall (1996), Panko and Halverson (1996), Panko (1998), Panko (2000), Rajalingham, Knight, and Chadwick (2000), and Lawrence and Lee (2004). The value of team-based code inspection is covered in McCormick (1983), Jones (1998), and Boehm and Basili (2001). Further discussion of spreadsheet design can be found in Nevison (1987), Raffensperger (2000), Rajalingham, Chadwick, Knight, and Edwards (2000), Grossman (2002), Colver (2004), and O'Beirne (2005). The role of the spreadsheet in financial institutions is discussed in Croll (2005). Stroustrup (1997) describes the principles underlying the popular programming language C++.

Boehm, B. and Basili, V.R. (2001). Software Defect Reduction Top 10. *Computer* 135–137.

Colver, D. (2004). Spreadsheet Good Practice: Is There Any Such Thing? Proceedings of the 5th Annual Conference of the European Spreadsheet Risks Interest Group. Klagenfurt, Austria, July.

Croll, G.J. (2005). The Importance and Criticality of Spreadsheets in the City of London.

Grossman, T. (2002). Spreadsheet Engineering: A Research Framework. Proceedings of EuSpRIG 2002 Symposium. University of Wales Institute. Cardiff, UK, July.

Hall, M.J.J. (1996). A Risk and Control Oriented Study of the Practices of Spreadsheet Application Developers. Proceedings of the Twenty-Ninth Hawaii International Conference on System Sciences. Kihei, Maui.

Jones, T. C. (1998). *Estimating Software Costs*. New York: McGraw-Hill.

Kavanagh, J. 1997. Shoddy Business Models Breed Financial Disaster. *Computer Weekly* 19 June.

Lawrence, L. and Lee, J. (2004). Financial Modelling of Project Financing Transactions. Presented to the Institute of Actuaries of Australia, Financial Services Forum. Sydney, Australia: The Institute of Actuaries of Australia.

McCormick, K. (1983). Results of Code Inspection for the AT&T ICIS Project. Paper presented at the Second Annual Symposium on EDP Quality Assurance.

Nevison, J.M. (1987). *The Elements of Spreadsheet Style*. Upper Saddle River, NJ: Prentice Hall.

O'Beirne, P. (2005). *Spreadsheet Check and Control*. County Wexford, Ireland: Systems Publishing.

Office of Inspector General, U.S. Department of Education. (2003). Audit of the Colorado Student Loan Program's Establishment and Use of Federal and Operating Funds for the Federal Family Education Loan Program. Region VII-Kansas City. http://www2.ed.gov/about/offices/list/oig/auditreports/a07c0009.pdf.

Panko R.R. (1988). *End User Computing: Management, Applications, and Technology*. New York: Wiley.

————. (1998). What We Know About Spreadsheet Errors. *Journal of End User Computing* 10(2):15–21.

————. (2000). Two Corpuses of Spreadsheet Error. Proceedings of the Thirty-Third Hawaii International Conference on System Sciences. Maui, Hawaii.

Panko, R. and Halverson, R. (1996). Spreadsheets on Trial: A Survey of Research on Spreadsheet Risks. Twenty-Ninth Hawaii International Conference on System Sciences 2:326–335, Kihei, Maui, Hawaii. January.

Raffensperger, J.F, (2000). The New Guidelines for Writing Spreadsheets. Department of Management, University of Canterbury, Christchurch, New Zealand.

Rajalingham, K, Knight, B., and Chadwick, D. (2000). Classification of Spreadsheet Errors. In *Spreadsheet Risks, Audit and Development Methods*, vol. 1. EuSpRIG, University of Greenwich.

Rajalingham, K., Chadwick, D., Knight, B., and Edwards, D. (2000). Quality Control in Spreadsheets: A Software Engineering-Based Approach to Spreadsheet Development. Proceedings of the Thirty-Third Hawaii International Conference on System Sciences. Maui, Hawaii.

Santayana, G. (1925). *The Life of Reason, or, The Phases of Human Progress: Reason in Science*. New York, New York: Charles Scribner's Sons.

Stroustrup, B. (1997). *The C++ Programming Language*, 3rd ed. Boston, MA: Addison-Wesley Longman Publishing Inc.

Taylor, M.J., Moynihan, E.P., and Wood-Harper, A.T. (1998). End-User Computing and Information Systems Methodologies. *Information Systems Journal* 8:85–96.

Endnotes

1. The programming language C++ was developed in the early 1980s at Bell Laboratories. It was based upon the earlier programming language known as C. One of its key objectives was to make writing good programs easier and more pleasant for the computer programmers. For further details on the language, see Stroustrup (1997).

2. Why are they so happy? Well, on Big City wages with no end in sight for the completion of their project, and it growing more complex by the day, wouldn't you be happy?

3. See, for example, the extensive survey by Croll (2005) on the role of spreadsheets in financial and other institutions in the city of London.

4. See Taylor, Moynihan, and Wood-Harper (1998).

5. Spreadsheet development is inherently different from conventional software development because spreadsheet programmers are often end users. See, for example, Croll (2005) who reports that people who create or modify spreadsheets in the financial markets are almost entirely self-taught.

6. The European Spreadsheet Risks Interest Group (www.eusprig.org), founded in March 1999, serves as a focal point for bringing together academics, professional bodies, and industry practitioners throughout Europe to address the issues surrounding spreadsheet integrity.

7. See the European Spreadsheet Risks Interest Group Web site (www.eusprig.org) for details of other spreadsheet disasters.
8. Details of the rapidly contrived conference call to sooth investor nerves is available at TransAlta (http://www.transalta.com/).
9. HealthSouth is one of the largest healthcare services providers in the U.S. The founder and former chief executive officer, Richard Scrushy, made history by being the very first CEO charged with violating the 2002 Sarbanes-Oxley Act. Fifteen other former executives pleaded guilty to various malfeasance and malpractice charges. Rather unusually five former chief financial officers testified against Scrushy. Scrushy was acquitted of all charges. Hannibal "Sonny" Crumpler—the former controller of the company's outpatient rehabilitation division, was not so fortunate. He was found guilty by a jury and convicted of conspiracy and making false statements to auditors. For more information on this fascinating case see *CFO Magazine* (http://www.cfo.com/article. cfm/5191992/c_5192704?f=archives&origin=archive).
10. Quote from SmartPros (http://accounting.smartpros.com/x48253.xml).
11. Quote from Miracle Solutions (http://www.auditexcel.co.za/war.html#8).
12. Quote from The Risks Digest (http://catless.ncl.ac.uk/Risks/16.72.html%20 #subj1).
13. The European Spreadsheet Risks Interest Group offers more details. On their site they report: "The National Australia Bank wrote down the value of its US mortgage business HomeSide Lending by a massive AUS$3 billion. The news triggered a free fall in the NAB's share price that knocked more than $6.5 billion of the bank's market value. Contributing to the write down was an incorrect interest rate assumption fed into HomeSide's financial modelling. This alone has cost the lender's $755 million. A selling spree knocked more than 13 per cent of the value of NAB shares." See European Spreadsheet Risks Interest Group (http://www.eusprig.org/stories.htm).
14. Quote from Miracle Solutions (http://www.auditexcel.co.za/war.html#8).
15. On February 5, 2002, AIB, one of Ireland's largest banks, announced that its American subsidiary, Allfirst, had incurred losses of around $700 million. This, at the time, was Ireland's largest ever financial scandal. It transpired that John Rusnak, a currency trader in the Baltimore office, had been conducting unauthorized and losing trades in Japanese Yen.
16. Quote from European Spreadsheet Risks Interest Group (http://www.eusprig. org/stories.htm).
17. Quote from the Office of Inspector General (2003).
18. One benefit of this process is detailed knowledge about how much of a corporation's financial reporting is sensitive to spreadsheet risk. Many corporations will have found much of their financial reporting is generated from spreadsheet models and that these spreadsheets may not have been well controlled or formally designed, developed, and thoroughly tested.
19. Looked at the 30 most financially significant spreadsheets used for project financing in an Australian consulting firm.
20. Furthermore, it would allow programmers to compile a preexisting spreadsheet into a stand-alone package that could run on computers that do not have a spreadsheet application or used as a subroutine or function from C++, FORTRAN, Java, or any other programming language.

21. See for example *Visual Baler,* which is a windows spreadsheet for creating stand-alone business applications. Another application is *Modelmaster/Excelsior,* a spreadsheet language, which can be used to create compilable spreadsheets.
22. For example, see Rajalingham, Chadwick, Knight, and Edwards (2000).
23. See, for example, Nevison (1987) or O'Beirne (2005) .
24. For example, see Colver (2004).
25. For example, see Panko (1998).
26. Raffensperger is a critique of software engineering prescriptions for spreadsheet design which specify modular separation of data, calculation, and output, in other words our Rule 2 which advocates isolation. The essence of his argument against isolation is that writing a spreadsheet is not like writing a computer program. Users, for the most part, do not want and will not use software engineering approaches. While this may be partly true, it is also true (and as we have seen) that spreadsheet development without discipline can lead to serious errors. Where IT departments (or others) are involved in the development of mission critical spreadsheets, best practice dictates isolation.
27. For discussion and empirical evidence supporting the notion that team code inspection can detect more errors than individual inspection see, for example, McCormick (1983), Jones (1998), and Boehm and Basili (2001).
28. For further discussion see the two articles by Tom Groenfeldt, the first entitled, Why In-House Systems Fail, and the second, Why Vendors Have An Edge. See both at DerivativeStrategy (http://www.derivativesstrategy.com).
29. The European Spreadsheet Risks Interest Group consists of academia and industry promoting research regarding the extent and nature of spreadsheet risks, methods of prevention, and detection of errors.

9

How to Guarantee Success by Understanding the Nature of Failure

You have been selected to lead a committee whose remit is to procure a new risk management system to be integrated into your company's existing informational systems infrastructure. Browsing through professional risk management publications, you cannot help but notice the long list of vendors offering risk management solutions.[1] This situation presents a great opportunity for selecting a solution that fits the exact requirements of your business. It also comes with problems associated with choosing from such a wide variety of systems. There is a degree of career risk involved, which arises because of the resources and time allocated, cost of the system, and unforeseen implementation and use issues. The success of the project, and how it reflects on you and the selection committee, could depend in large part on choosing the right vendor.

Now you and the members of the committee are about to leap into the fray of competing vendors, all of whom say they can provide the right risk management tools, integrate them into your preexisting information systems, and provide dedicated support down the road. Like Roald Dahl's Charlie in Willy Wonka's chocolate factory, it all sounds simply too delightful:

> Mr. Willy Wonka can make marshmallows that taste of violets, and rich caramels that change colour every ten seconds as you suck them, and little feathery sweets that melt away deliciously the moment you put them between your lips. He can make chewing-gum that never loses its taste, and sugar balloons that you can blow up to enormous sizes before you pop them with a pin and gobble them up. And, by a most secret method, he can make lovely blue birds' eggs with black spots on them, and when you put one of these in your mouth, it gradually gets smaller and smaller until suddenly there is nothing left except a tiny little dark red sugary baby bird sitting on the tip of your tongue.[2]

While there is much to know about financial risk management, a quick look at the bookshelves in your local store or bookselling Web site may leave you feeling like Charlie in Willy Wonka's factory—a little bemused. Books on quantitative methods, valuation, and derivative pricing are the dominant themes. This literature puts a lot of thought into the tools of risk measurement and modeling. However, relatively little effort has gone into providing advice on those factors that increase the probability of successfully selecting and deploying the most appropriate risk management system.

Perhaps the subject is thought of as the purview of information systems specialists and project managers alone. Nothing could be further from the truth. Selecting an appropriate vendor to provide a risk management system—one that addresses the specific business risks of your organization—is an important step in developing an effective risk oversight mechanism. Unfortunately, a favorable outcome is not guaranteed. Improving the probability of a successful vendor selection process requires some understanding of the factors that contribute to both success and failure. This chapter discusses aspects of the vendor selection issue, drawing on lessons learned in the information systems and software engineering community.

The Value Added of Vendor Risk Information Systems

Firms competing in the financial services marketplace face a stark reality: continuously anticipate and vigorously respond to the evolving demands for products and services, or perish. In today's intensively competitive environment, with rapidly advancing internationalization and commoditization of both services and products, business strategy not only determines performance relative to peers, it also governs business survival. Modern business strategy generally requires aggressive and efficient use of information systems. Risk management is not immune from this trend. Rapid advances in risk management technology, a proliferation of vendors, and changes in the regulatory environment present senior management with both increased pressure and opportunities for the management of business, financial, and other risks.[3]

The idea of seamless integration of all risk information flowing through a large complex financial services corporation—from operations, equity, credit, fixed income, property, and overseas units—is very appealing, especially to corporate risk managers who have had to struggle with incompatible information systems, inconsistent risk terminology, and operating and risk measurement practices. It is also appealing to senior management who can use consolidated risk management information to better understand the risk-return trade-off between various business lines and, therefore, make more informed decisions regarding how to invest capital. Consolidated risk information also improves the ability to locate and exploit natural hedges and thus has the potential to yield a vital competitive advantage.[4]

One of the oft-overlooked benefits of implementing a firm-wide risk system is the role it can play in raising a broader, firm-wide risk consciousness, which can completely transform an entire corporate risk culture. Such systems can inject additional risk management discipline into the corporate

WHERE IS THE VALUE ADDED?

An integrated risk management information system can result in significant improvements to existing processes and, therefore, reductions in overall operational costs. This is because such a system offers the ability to:

1. Automate and integrate business risk processes.
2. Share common data and risk management practices across the entire enterprise.
3. Produce and access risk information in real time (or near real time).
4. Allow geographically dispersed and possibly autonomous business units to roll up standardized risk management information for corporate reporting.

structure by pushing business units toward standardization and thereby reducing the heavy drag on risk management productivity and performance that can often arise as a consequence of business-unit-specific risk terminology, operations, and risk measurement.

However, there are considerable technical hurdles involved in developing a risk management system which integrates different types of risk from disperse lines of business located in possibly differing regulatory jurisdictions. However, just as advances in financial engineering and the regulatory environment are changing the operating methods and business cultures of financial institutions, vendors[5] are developing new information systems to capitalize on these developments.

In very many cases, the long-term productivity and connectivity gains from the flow of timely risk management information created by a vendor-supplied risk system are so compelling that not adopting one is out of the question. Even when such gains are excluded, rising demand for vendor-supplied risk management systems continues to be driven by regulators,[6] rating agencies, and boards of directors[7]—who are increasingly calling for more transparency and measurement surrounding risk-related issues.

How to Guarantee Success by Understanding the Nature of Failure

Selection of an appropriate vendor and successful implementation is challenging and often elusive. Failure, although never an option, is all too often the reality of large-scale information systems projects. Dove (2004) provides a useful categorization for benchmarking project failure:

> Project failure can be defined under two basic categories:
>
> 1. A project that consumes resources but fails to deliver an acceptable ROI.
>   1. The project is terminated before completion.
>   i. Needs cease to exist—the world changed unpredictably.
>   ii. Necessary resources become insufficient or unavailable.
>   iii. Decision makers have a change of heart or are replaced by ones who don't care.
>   2. The project was ill defined so resources were inefficiently applied as it developed definition.
>   3. The project was incorrectly defined, resulting in user rejection or insufficient value.
> 2. A project that consumes resources but fails to deliver as proposed.
>   1. The project exceeds budget.
>   2. The project exceeds time.
> 3. The project doesn't meet spec.

Depending upon which academic study you read, between 70% to 90% of large information systems projects fall into Dove's second category, the failure rate is anywhere between 50% to 80% with 20% to 40% stopped while they are underway.[8] Truly *eye watering* failures can be found in both government sponsored systems and the corporate sector. Three years after the September 11, 2001 terrorist attacks on the United States of America and following multiple missed deadlines and a price tag of around $170 million, the Federal Bureau of Investigation's new case management system was deemed unfit for purpose and scrapped. The corporate sector is not without its fair share of fiascos; for example, in 1993, Taurus, the London Stock Exchange's planned automated transaction settlement system, was ditched after 5 years of error-ridden software development. *Wired Magazine*[9] reflecting the *zeitgeist* famously reported:

> Britain's had a bad year in information technology. In March, London's Stock Exchange abandoned its partially completed Taurus system to administer share-trading, despite having already spent well over $500 million on it. Traditionally, British managers have shared James Bond's attitude toward technology—a potentially wonderful but dispensable luxury. Real men don't need to think too hard about widgets (technology too complicated to take seriously). If the boffin's (person who take widgets seriously) laser-guided wristwatch doesn't get the baddies, a swift karate kick will always suffice.

Davenport (1998), writing 5 years after the Taurus fiasco, makes clear, information systems project failures continue to have a significant impact on the business environment:

The growing number of horror stories about failed or out-of-control projects should certainly give managers pause. FoxMeyer Drug argues that its system helped drive it into bankruptcy. Mobil Europe spent hundreds of millions of dollars on its system only to abandon it when its merger partner objected. Dell Computer found that its system would not fit its new, decentralized management model. Applied Materials gave up on its system when it found itself overwhelmed by the organizational changes involved. Dow Chemical spent seven years and close to half a billion dollars implementing a mainframe-based enterprise system; now it has decided to start over again on a client-server version.

Apart from wasting precious financial resources[10] and causing a great deal of disruption, failure can result in the loss of goodwill between employees and business lines. It may also (probably will) reflect poorly on you as the manager or member of the risk system selection committee. Although risk management technology projects falter for a variety of reasons, when a project does fail, senior management and executives associated with the project may pay dearly since the sunk resource costs may be high, and a failed project may hinder the effective risk management of the business for many years to come.

Developing a Winning Game Plan

Given the corporate and personal risk involved, choosing a risk system vendor can be an intimidating, even overwhelming experience—it does not have to be. As the words of the French vagabond, dreamer, and devotee of pleasure, Jean de La Fontaine[11] underscored, success at any endeavor begins with a plan:

> NOT quite so fast, rejoined our smart gallant,
> First know the plan, before consent you grant;
> There is an ill attends the whole affair;
> But what below, alas! is free from care;
> This juice, possessing virtues so divine,
> Has also pow'rs that prove the most malign:
> Whoe'er receives the patient's first embrace;
> Too fatally the dire effects will trace;
> Death oft succeeds the momentary joy;
> We scarcely good can find without alloy.

At the onset, clarity of purpose, understanding of the business implications, and the expectation of measurable benefits, provide an important benchmark against which the selection process should be referenced. This is the very same lesson emphasized by Davenport (1998) in a comprehensive study of enterprise systems[12]:

> In fact, having now studied more than 50 businesses with enterprise systems, I can say with some confidence that the companies deriving the greatest benefits from their systems are those that, from the start, viewed them primarily in strategic and organizational terms. They stressed the enterprise, not the system ... the companies that have the biggest problems—the kind of problems that can lead to an outright disaster—are those that install an ES [Enterprise System] without thinking through its full business implications.

Therefore, like any other endeavor in which we seek success, the struggle to winnow down the vendors to the right one or two should not be entered into without a clear, focused vision, underpinned by a solid business case[13] and plan. *Always remember technology will not save you, a great strategy executed well will!* For maximum probability of success, the vendor selection project should be aligned with the business goals detailed in the corporate strategic plan. Expectations about the strategic benefits, resources, costs, risks, and timeline should be used to underpin the central objective of finding and selecting an appropriate vendor. Without a clear vision aligned to the objectives of the corporate strategic plan, do not be surprised when the risk vendor selection project you have wholeheartedly sponsored and supported, that you sincerely believed would fly is perceived as having little "value added." If the project falls into this status, it may be more easily canceled or team members and other resources pulled, causing the project to stall and fall clumsily back to earth. Never forget the project management adage:

> If you fail to plan, you must plan to fail!

Creating a High Performance Team

One of the first steps in choosing a new risk system from the list of competing vendors is to assemble together a vendor selection and assessment team. It sounds simple, yet it is surprising the number of technological projects that are initiated with team members who are confused by technology, computationally challenged, strongly influenced by personal interests or preconceived biased perceptions, or with very limited time to devote to the vendor selection and assessment process. Such teams will inevitably fail to function optimally because they do not have the appropriate technical, project management skills, or time commitment. Getting the right people on board hinges heavily on crafting together a high performance team; a vendor selection committee which can define necessary requirements and maintain a clear sense of purpose, willing to invest large amounts of time (as necessary) to the vendor selection process, with the ability to develop and express deep collective and individual feelings about the proposed risk system, its purposes, and its future evolution. Such teams are able to focus on the issues and variables that can make a real difference to the eventual outcome. High performance teams are characterized by *buy in* by all team members about the team mission.

Team direction and motivation has long been associated with the strength of shared values within the team.[14] This is why, at the outset, a clear, focused vision, underpinned by a solid business case and plan are essential. Without this, the team may eventually lose focus, disintegrate, or otherwise through internal bickering become entirely ineffective.[14] For a team member to buy in demands a set of shared values with other members of the team—be it contribution, determination, altruism, loyalty, or any other value.

The members of the team should be selected from subject matter experts and users across the relevant business units. Accept a priori that a selection team made up of well-paid senior people with the right specialized skills is worth far more per dollar to your organization than a group of lower-cost junior individuals. A team constructed of junior individuals will almost inevitably require more time as they fumble sluggishly through the assessment and selection process. Indeed, there is a growing body of academic evidence to suggest that team members should be the best people in the organization.[15] This is because sustained high performance depends heavily on the temperament and intellectual caliber of the individual team members. The best people in an organization are more likely to understand and be able to explain new concepts and processes. In addition, such individuals tend to have the requisite business and technical knowledge surrounding existing business processes and the requirements that will need to be satisfied by the system.

Since both business and technical knowledge are essential for success, effective involvement of a range of key individuals from across different business units is required; different perspectives can provide a source of experience and innovative thinking, which may enhance team performance. Active participation across business lines can have the effect of inducing clarity and consensus concerning the system requirements. In a high performance team,[16] every member is an essential contributor, and leadership encourages a balanced high-energy dialog where every member is contributing appropriately.

The Important Lesson of $\frac{1}{2} \times n \times (n - 1)$

In one of the most influential computing books of the 20th century, Brooks (1995) observed that a team with n individuals has $\frac{1}{2} \times n \times (n - 1)$ communication links. Why is this important? Because as Elenbaas (2000) states:

THE BOTTOM LINE ON TEAM SIZE

Clearly, one needs to keep a keen eye on the size of the team as size will also be a factor in how long it takes to complete the vendor assessments. There is often pressure from management to add more people to a team in the belief that the overall vendor assessment can be completed faster. Such pressure should be resisted if the consequences result in an unwieldy group. Team size, therefore, needs to closely monitored. Too big can create too many opinions, too small and you risk bias from incomplete viewpoints.

> Projects are about communication, communication, communication.

A team of 15 has almost 4 times more communication links than a team of 8. Each communication link requires time for maintenance. The higher the number of links, the greater the number of conversational threads and the more difficult is the management of coordinated communication. Individual team members may become frustrated with their inability to express themselves with the whole group so they interact with a subset of the group and the team fragments. This, in turn, increases the potential for miscommunication with all its attendant problems.

In both the academic and practitioner literature, smaller teams have been long recognized as having an edge over larger teams, at least in the case of information systems projects.[17] This is not because a small team is advantageous, in and of itself, but that the alternative—a large team—is so disadvantageous, primarily because of the burden of maintaining the communication links between the team members. Various empirical studies[18] have revealed that a project team should be just large enough to do the work it is required to do. Any smaller and the team may not be technically able to perform its tasks. This is because team members are likely to be so consumed that common goal setting, communicating, and becoming a cohesive unit have to be neglected. In the wider business literature, a team size, which is just large enough, is captured by the principle of least group size.[19] A team that breaches this principle, in terms of having too many members, will produce process losses due to increasing coordination and communication requirements. Overly large teams are characterized by significant levels of *social loafing*, a term used to describe the reduced effort of the individual team member.[20]

The Critical Role of Executive Buy-In

The plain and simple fact of the matter is that vendor choice can affect company profitability and should therefore be everyone's business, especially

ARE KEY DECISION MAKERS ON BOARD?

It is critical to garner the backing and commitment from key decision makers, those who control organizational priorities, strategy, commit funds and resources. In practice, this requires at a minimum the following:

1. Public and explicit statement from senior decision makers that the vendor selection project is a top priority.
2. Gathering onto the team senior decision makers.
3. That key decision makers remain actively engaged throughout the entire vendor assessment and selection process.

management and senior executives. If senior management places little value on a new risk system, the vendor selection project is likely to be seen as not vital to the critical mission of the organization. Vendor selection projects that are delegated down to line level managers without strong executive support tend to drift along, stop when problems arise or else dissipate in fragmented directions. Such projects are unlikely to garner top management involvement and support, and are thus less likely to be institutionally sustainable in the long run; indeed, it may not survive its first major "crisis." Such projects, therefore, have a high risk of failure.[21]

Empirical study after empirical study has shown that top management support is needed throughout the vendor selection and implementation process.[21] When there is an executive sponsor sitting in status meetings, reviewing plans, meeting with team members, and taking part in vendor assessments, the team is more likely to remain focused on the project objectives and obstructions are removed more swiftly.

Clarifying Your Requirements

With a high performance team in place and buy in from senior executives and team members, the process of requirements elicitation can begin. At this stage, the selection team should produce a statement specifying:

- The nature of the risk information required;
- For whom it is required;
- And, the required timeliness.

For example, in the case of a single equity portfolio, this statement might specify that the calculation of value at risk, tracking error, and stress testing be provided on a monthly basis to the head of equities and the portfolio manager. In the case of multiple portfolios across many asset classes for an entire financial institution, each broad class of user (risk, portfolio and business managers, traders and corporate headquarters, and so on) should be identified (alongside named individual users from each of these areas) and their risk informational requirements.

This process is time intensive and will generally require members of the selection committee to discuss formally (say through structured interviews) or informally with senior executives, department management, and end users. Gathering and writing down these requirements can seem like an overwhelming task, and one might feel that time would be better spent looking directly at the vendors. However, gathering together a clear set of risk requirements at an early stage aids project transparency and concentrates team focus. The identified risk information (value at risk [VaR], stress testing, cash flows, instrument valuation, credit exposure, and so on) should be mapped to each area alongside the timeliness requirements. We term such a mapping, a Risk Information Requirements Table (RIRT) (see a RIRT illustrated in Table 9.1).

Using the RIRT as a guide, the selection team can begin production of the Risk System Requirements Documentation (RSRD). RSRD is a comprehensive list of the necessary features required by the risk system. The elicitation process will involve team members taking a critical look at the features of their current risk system(s) and identifying all current shortcomings. Table 9.2 provides an illustration of what an RSRD might look like. The process of producing an RSRD is useful because it will highlight that functionally of the old system(s), which you wish to retain and the additional functionality that is required by the new vendor system.

TABLE 9.1

Sample Risk Information Requirements

Area	Information	Timeliness
Equities:	VaR	*Daily*
Derek Wong—Portfolio Manager	Stress Testing	Monthly
Shola Beecher—Business Head	Volatility	Weekly
Bernard Agbaje—Risk Manager	Derivative Pricing	Daily
Risk Group:	*VaR*	*Monthly*
John Okube—Analyst	Expected Shortfall	Monthly
	Credit Exposure	Monthly
...	...	...
Corporate:	*Credit Exposure*	*Quarterly*
Terry Andrews—Risk Analyst	VaR	Quarterly
Wendi Chow—Business Audit	Cashflow Maps	Monthly
	Duration Maps	Monthly
	Present Value Reports	Monthly

TABLE 9.2

Example of a Risk Systems Requirement Document with Prioritization

Point	Requirements	Priority Level	Item 1	Item 2	Item 3	Item 4	Item 5
1	Derivatives Coverage	Critical	Equity	Fixed Income	Currency	Energy	Real Estate
2	Fixed Income	Critical	G8 Treasuries	MBS	ABS	CMBS	Euro Dollar Instruments
.	.	.	.	.	.	.	.
.	.	.	.	.	.	.	.
.	.	.	.	.	.	.	.
248	Statistical Modeling	Desirable	Principal Components	Parametric	Nonparametric	Econometric	Time Series Modeling
249	Value at Risk	Critical	Var-Covar	Historical Simulation	Stochastic Simulation	Combo	Customization
250	Risk Reports	Important	Customization	Limit Setting and Monitoring	Hierarchy Risk Analysis	Filtering	Screening
251	Usability	Important	Ease of Use	Customizability	Syntax	Menu Driven	Programmability
252	Accounting	Critical	FASB133	GAAP	Sarbanes-Oxley	Statutory	Customization

MY VENDOR SELECTION TEAM TELLS ME THEY ALREADY KNOW WHAT IS REQUIRED, SHOULD I LET THEM SKIP FORMAL ELICITATION?

The objective of this stage of elicitation is to establish at the outset a reasonably stable requirements baseline. Poorly defined business requirements are a leading cause for project failure. If the new requirements are well known, at the very least the selection team should be able to produce a detailed RIRT and RSRD. The RIRT will ensure that it is clear to you, the team, and your organization what information is required, for whom, and its timeliness. You should review it. Obvious omissions are a sign that the process may be moving in the wrong direction. Similarly, the RSRD should be made available to all interested parties. Any omissions will soon be spotted. Other benefits of taking time to complete this step thoroughly include:

1. The RSRD might only highlight a few areas of weakness, which may be fixed via additional purchases or modules rather than an entire new system.
2. A thoroughly constructed RIRT and RSRD will provide further ammunition with which to lock in support from business areas and senior management.
3. The process might bring to light additional inefficiencies in the current risk systems processes which can be eliminated, thus saving costs and increasing productivity.

Common areas covered in the RSRD include financial instruments, pricing models, statistical techniques, risk reporting, accounting, and so on. Since there are many capabilities supported by some, but not all, vendor risk management systems, it is also important to prioritize the identified requirements, for example, "Critical" for essential requirements, "Important" for the next level, "Desirable," and so on. When working through the requirements elicitation process, it is helpful to remind yourself and the team that the system will be in place for a long time. As such, it is beneficial to take a forward-looking approach toward defining and prioritizing requirements.

Once the RSRD has been agreed upon and signed off by the selection committee, it can be used to form the basis of a Request For Proposal (RFP).[22] An RFP is a document of the risk system requirements a vendor system must satisfy in order to win your business. An effective RFP requires a considerable amount of work. Typically, it will consist, at a minimum, of the following areas:

1. Background information.
2. Scope of required services.

3. Threshold criteria.
4. Vendor company profile.
5. Vendor system details.
6. Vendor response criteria.

The background information is essentially an executive summary containing a brief description of the project. The actual content varies from organization to organization. At the very minimum, it will also need to contain further information about your organization, the investment systems and risk management operating environment, and details of risk management functions and responsibilities from an information systems perspective.

The purpose of providing a threshold criteria is to ensure that all vendors who respond to your RFP meet a minimum level before they are considered any further. For example, your organization might feel uncomfortable working with a vendor provider who has less than 3 years experience, those who have no prior knowledge customizing their risk systems to your specific line of business, or vendors without primary headquarters or a significant presence inside your particular country or region of operation. The criteria should be chosen such that it allows your selection committee to concentrate only on those vendors capable of implementing and supporting the new system in a cost effective and administratively efficient manner. Thus, the threshold criteria acts as a self-selecting filter to weed out the no hopers from the pack.

The responding vendors should also supply a comprehensive company profile in response to specific questions developed by the selection committee. These questions are designed to elicit deep and meaningful insight into potential vendors. It is important that the vendor demonstrate its product is equipped with the functionality you require. When thinking about their response, ask yourself "How will their system consolidate behind the scenes?" A modern risk system should allow for distributed users, from across the office to across the world, to enter data and have it all consolidated in one place. The system should be able to bring everything together smoothly and efficiently so that once the data has been entered, the risk manager, analyst, or senior executive can easily run a report. It is equally important to select a vendor who you sense is accompanied by a strong team that can assist the implementation and provide ongoing support.

The Truth about Project Managers

Project management is about delivering a project in line with a client's expectations in terms of quality, cost, and time. As we have already seen, there is considerable evidence that a majority of information technology projects fail to meet any of these objectives. Almost all of these failed projects have project managers:

> There is little doubt that project management is generally not delivering the results it promises. The Standish Group's 1994 Chaos Report finding that only 16% of software projects are completed on-time and on-budget has been widely cited. Robbins-Gioia Inc. conducted a similar, although smaller, study of construction projects in which 44% of participants reported projects with cost overruns of 10% to 40%. And Terry Cooke-Davies analyzed 136 (mainly) European projects executed between 1994 and 2000 and found that the mean performance against budget was a 4% cost escalation while mean schedule performance was 16% late. ... Project managers will argue these depressing results represent not a failure of project management, but a failure to apply project management effectively. This argument certainly has intuitive merit. But the evidence supporting this claim is scant and unconvincing.[23]

The status of project management, never strong (even in the good years), reached an all-time low in the winter of 2000, when it became the official whipping boy in a parliamentary debate ostensibly about the sale of the United Kingdom's National Air Traffic Services (NATS):

> The issue here is the failure of project management—we understand that. That is why we want to look for solutions for project management rather than simply selling off NATS. The argument is that project management in the public sector is particularly weak. Yet the channel tunnel was hardly a wonderful example of project management in the private sector. Lord Macdonald argues that there needs to be a shareholder. It did not work with the channel tunnel. There was a single shareholder in the dome, and it did not work there. The argument does not hold water. (John McDonnell, November 15, 2000, House of Commons, UK)

Despite over five decades of project management evolution, comparatively few projects are judged to be wholly successful. The repeated failure of project management practices in a wide variety of settings have led some to question the value of the discipline.[24] Yet, although much maligned, project managers can play a key role in the facilitation of team meetings, channeling productive ideas, conflict resolution, and ensuring frequent, formal, well-planned team communication. It is not necessary that the project manager be expert in the field of risk management technology, rather the key requirement is that he or she is experienced:

> Corporate America spends more than $275 billion each year on approximately 200,000 applications software development projects. Many of these projects will fail, but not for lack of money or technology; most will fail for lack of skilled project management. The Standish Group (1999).

HOW CAN I TELL IF OUR VENDOR ASSESSMENT AND SELECTION TEAM WILL BE HIGH PERFORMANCE?

Careful construction of a high performance vendor selection and assessment team can pay dividends later down the line in terms of improved competitive corporate performance via more efficient risk management. Such teams are characterized by skilled collaborative team members willing to work with people of different styles and business backgrounds with mutual respect. They operate in a collaborative environment with committed senior executive support and effective facilitation via project managers. High performance teams have been studied extensively in the academic literature.[26] Corporate teams that excel have a number of common characteristics:

- Commitment to purpose.
- Designated roles.
- Near-term objectives.
- Teamwork focused on the task-at-hand.
- Strong and clear leadership.
- Individual and collective accountability.
- Executive buy in.
- Experienced project management skills.

Recently there has emerged a body of evidence suggesting a positive correlation between the value created by using a project manager and their level of project management experience.[25] The larger the project, the more need there is for experienced people with excellent planning, oversight, organization, and communications skills, in other words an experienced project manager. An experienced project manager may get above-average results from average teams, whereas great employees can have much of their potential squandered by mediocre project management.

For Further Thought

Selecting a risk systems vendor is, in many senses, the first step in a process that should end with a fully functioning integrated risk system. Successfully negotiating the advertising and sales pitches to identify those vendors who are capable of satisfying your business requirements will set a solid foundation on which successful implementation and operation can be founded. Moving beyond vendor selection[27] toward implementation

VENDOR SELECTION

Key Point 1: Modern business strategy requires aggressive use of information systems. Risk management is not immune from this trend.

Key Point 2: Understand the nature of failure—there are considerable technical challenges involved in developing a risk management system. Around 50% to 80% of large information systems projects fail with 20% to 40% stopped while they are in progress.

Key Point 3: The benefits can be considerable both in terms of improvements to existing processes and the quality of risk information provided to senior management.

Key Point 4: Clarity of purpose, understanding of the business implications, and the expectation of measurable benefits provide an important benchmark against which the vendor selection process should be referenced.

Key Point 5: Create a high performance team selected from subject senior experts and users across relevant business units.

Key Point 6: Team size is important. A team with n individuals has ½ × *n* × *(n − 1)* communication links. Each link requires maintenance. The more links, the higher the likelihood of a communication breakdown.

Key Point 7: Executive buy in is critical.

Key Point 8: Experience counts!

Key Point 9: An experienced project manager can get above-average results from average teams.

Key Point 10: Great employees can have much of their potential squandered by mediocre project management.

Key Point 11: High performance teams have a strong commitment to purpose with clearly designated roles. In addition, they have strong and clear leadership with executive buy in.

Key Point 12: Choosing a risk system vendor may be a critical step toward creating a fully functioning integrated risk management system.

and operation offers challenges around organizational structure, managing user expectations, change management, maintenance and support, and resource commitment. Unfortunately, adhering closely to the ideas presented in this chapter cannot entirely ensure the avoidance of failure; but at the very least you will be able to recognize a half-baked vendor selection project when you see one and steer your organization (or failing that yourself) away from it.

Unfortunately, there is no simple answer to delivering success. Awareness of the issues raised in this chapter will give you an edge. Questions that may require further thought include:

1. Are you confident you have clarity of purpose and understanding of the business requirements?

2. Can you identify a concise list of measurable benefits expected, which may act as a benchmark against which the selection process can be referenced?

3. What are the strategic and organizational imperatives? Do they dominate other considerations? Are they the primary drivers?

4. Poorly defined business requirements are a leading cause of project failure. Do you have at the very least a clear sense of:

   i. The nature of the risk information required,

   ii. For whom it is required,

   iii. And the required timeliness?

5. In a high performance team, every member is an essential contributor, and leadership encourages a balanced high-energy dialogue where every member is contributing appropriately. What is your perception of the capabilities and qualities of the selection team? Things to consider include:

   i. Technical knowledge.

   ii. Diversity across business units.

   iii. Personal interests.

   iv. Time availability.

   v. Sense of purpose and collective buy in.

   vi. Seniority and experience level of team members.

   vii. Collective and individual feelings about the proposed risk system, including preconceived perceptions.

6. Identify the shared values, which tie the team together. Things to consider include:

   i. Altruism

   ii. Loyalty

　　iii.　Contribution

　　iv.　Determination

7. Recall "The important lesson of $\frac{1}{2} \times n \times (n - 1)$." What value does it take for your team?

8. What value does senior management place on the system?

　　i.　Who is the executive sponsor?

　　ii.　What role will they play in the selection process?

　　iii.　Are other key decision makers engaged?

9. Who is responsible for project management?

　　i.　What is their level of experience?

10. What is your honest answer to each of the following about your team?

　　i.　Commitment to purpose.

　　ii.　Clearly designated roles.

　　iii.　Strong near-term objectives.

　　iv.　Teamwork focused on the task-at-hand.

　　v.　Strong and clear leadership.

　　vi.　Individual and collective accountability.

　　vii.　Executive buy in.

　　viii.　Experienced project management skills.

Additional Resources

Consolidated risk management systems have significant potential for making quantum leaps in productivity, increasing the ability to compete, and maintaining sustainable competitive advantage. However, much of that potential remains to be achieved. Hand (1989), Dawes and Worthington (1996), Service and Maddu (1999), and Palanisamy and Sushil (2003) discuss various aspects of the link between information systems and business strategy. Whittaker (1999), Browning (1993), Davenport (1998), and Dove (2004) address the nature of information systems failure. The role of high performance teams is outlined in Katz (1982a, 1982b), Vaill (1982), Keller (1986), Lamb (1985), Larson and LaFasto (1989), Keller (1992), Browning (1993), Whittaker (1999), Amabile et al. (2001), and Dove (2004). Details of the role and effectiveness of project management are discussed in Ibbs and Kwak (1997a, 1997b, 2000) and also Brown and Adams (2000).

Amabile, T.M., Patterson, C., Mueller, J., Wojcik, T., Odomirok, P.W., Marsh, M., and Kramer, S.J. (2001). Academic-Practitioner Collaboration in Management Research: A Case of Cross-Profession Collaboration. *Academy of Management Journal* (April) 44(2):418–432.

Bingi, P., Sharma, M.K., and Godla, J. (1999). Critical Issues Affecting an ERP Implementation. *Information Systems Management* 16(3):7–14.

Brooks, F.P. Jr. (1995). *The Mythical Man-Month*. Boston: Addison-Wesley.

Brown, A. and Adams, J. (2000). Measuring the Effect of Project Management on Construction Outputs: A New Approach. *International Journal of Project Management* 18:327–35.

Browning, J. (1993). Tech Troubles Threaten London Stock Market. *Wired Magazine* July/August.

Buckhout, S., Frey, E., and Nemec, J.Jr. (1999). Making ERP Succeed: Turning Fear into Promise. *IEEE Engineering Management Review* 27(3):116–23.

Buys, C.J. and Larson, K.L. (1979). Human Sympathy Groups. *Psychology Reports* 45:547–553.

Campion, M.A., Medsker, G.J., and Higgs, A.C. (1993). Relations between Work Group Characteristics and Effectiveness: Implications for Designing Effective Work Groups. *Personnel Psychology* 46(4):823–850.

Carmel, E. and Bird, B.J. (1997). Small Is Beautiful: A Study of Packaged Software Development Teams. *Journal of High Technology Management Research* 8:129–148.

Davenport, T.H. (1998). Putting the Enterprise into the Enterprise System. *Harvard Business Review* 76:121–131.

Dawes, J. and Worthington, S. (1996). Customer Information Systems and Competitive Advantage: A Case Study of a Top Ten Building Society. *International Journal of Bank Marketing* 14(4):36–44.

Dove, R. (2004). Decision Making, Value Propositioning, and Project Failures— Reality and Responsibility. Proceedings of International Council on Systems Engineering (INCOSE) 2004 Region II Conference. (September):1–9.

Fria, R. (2005). *Successful RFPs in Construction*. New York: McGraw-Hill Professional Publishing.

Hand, M. (1989). Managing Strategic Investment in Information Systems. *Management Accounting* UK 67(9):46.

Harkins, S.G. and Petty, R.E. (1982). Effects of Task Difficulty and Task Uniqueness on Social Loafing. *Journal of Personality and Social Psychology* 43(6):1214–1229.

Ibbs, C.W. and Kwak, Y.H. (1997a). *The Benefits of Project Management: Financial and Organizational Rewards to Corporations*. Newton Square, PA: Project Management Institute Educational Foundation.

———. (1997b). Financial and Organizational Impacts of Project Management. Proceedings of the 28th Annual PMI Seminars & Symposium. Chicago. 496–500, September.

———. (2000). Assessing Project Management Maturity. *Project Management Journal* (31):32–43.

Karau, S.J. and Williams, K.D. (1993). Social Loafing: A Meta-Analytic Review and Theoretical Integration. *Journal of Personality and Social Psychology* 6(4):681–706.

Katz, R. (1982a). High Performance Research Teams. *The Wharton Magazine* Spring:29–34.

———. (1982b). The Effects of Group Longevity on Project Communication and Performance. *Administrative Science Quarterly* 27(1)(March):81–104.

Keller, R.T. (1986). Predictors of the Performance of Project Groups in R&D Organizations. *Academy of Management Journal* 29(4):715–726.

———. (1992). Transformational Leadership and the Performance of Research and Development Project Groups. *Journal of Management* 18(3) (September):489–501.

Nah, F.F. and Lau, J.L. (2001). Critical Factors for Successful Implementation of Enterprise Systems. *Business Process Management* 7(3):285–296.

Lamb, W. (1985). Building Balanced Innovation Teams. *Journal of Product Innovation Management* 2(2) (June):93–100.

Larson, C. and LaFasto, F.M. (1989). *Teamwork: What Must Go Right/What Can Go Wrong.* Newbury Park, CA: Sage.

Porter-Roth, B. and Young, R. (2001). *Request for Proposal: A Guide to Effective RFP Development* (Paperback). Addison-Wesley Professional.

Rosario, J.G. (2000). On the Leading Edge: Critical Success Factors in ERP Implementation Projects. Philippines: *Business World.*

Service, R.W. and Maddu, H. (1999). Building Competitive Advantage through Information Systems: The Organizational Information Quotient. *Journal of Information Science* 25(1).

Shaw, J. (2005). Managing All of Your Enterprise's Risks. *Risk Management Magazine* (September). Risk and Insurance Management Society.

Somers, T. and Nelson, K. (2001). The Impacts of Critical Success Factors Across the Stages of Enterprise Resource Planning Implementations. Proceedings of the 34th Hawaii International Conference of System Sciences. Maui, HI. 1–10.

Standish Group International. (1999). *Chaos: A Recipe for Success.* Boston, MA.

Sushil, P.R. (2003). Achieving Organizational Flexibility and Competitive Advantage Through Information Systems Flexibility: A Path Analytic Study. *Journal of Information and Knowledge Management* 2(3).

Thelen, H.A. (1949). Group Dynamics in Instruction: Principle of Least Group Size. *The School Review* (March):139–148.

Vaill, P.B. (1982). The Purposing of High-Performing Systems. New York, NY: *Organizational Dynamics* Autumn 11(2):23–40.

Wasserman, S. and Faust, K. (1994). *Social Network Analysis: Methods and Applications.* Cambridge, UK: Cambridge University Press.

Whittaker, B. (1999). What Went Wrong? Unsuccessful Information Technology Projects. *Information Management & Computer Security* 7:23–29.

Willcocks, L. and Sykes, R. (2000). The Role of the CIO and IT Function in ERP. *Communications of the ACM* (April).

Ziller, R.C. (1957). Group Size: A Determinant of the Quality and Stability of Group Decisions. *Sociometry* 20:165–173.

Endnotes

1. Increasingly, professional risk magazines and journals are publishing vendor surveys. For example, *Risk Magazine* (www.risk.net) publishes occasional vendor technology reports.
2. Dahl, R. (1964). *Charlie and the Chocolate Factory* (New York: Alfred A. Knopf).

3. For example, in the United States, the Gramm-Leach-Bliley Act of 1999 strength-ened integration in the financial services industry by creating a new type of bank holding company known as a financial holding company. Financial holding companies were permitted to engage in a range of financial activities, including insurance, securities underwriting and agency activities, merchant banking, and insurance company portfolio investment activities.

While the act strengthened financial integration between the banking and life insurance industries it also opened up competition among banks, securi-ties companies, and insurance companies. Financial holding companies, with their diverse financial activities, present challenges to consolidate risk manage-ment information. The traditional approach of managing risks in separate silos is inappropriate. This is because financial holding companies risks can be highly interdependent, making their efficient management at the business unit level difficult.

4. See Shaw (2005) who discusses the consequences of failure to monitor consoli-dated risks at the Ford Motor Company.

5. And also financial institutions themselves. This raises the interesting question of whether to develop the risk system internally or turn to an external vendor. The answer depends on the balance between the cost of managing the project internally versus the transaction costs associated with purchasing an external system. This chapter assumes the transaction costs are such that purchasing an external system is the optimal choice. Nevertheless, many of the lessons dis-cussed are equally applicable to internally developed risk systems.

6. Key legislative and regulatory initiatives in the U.S. include the Sarbanes-Oxley Act of 2002, the Patriot Act of 2001, and the Gramm-Leach-Bliley Financial Services Modernization Act of 1999. Other regulatory initiatives include the New Basel Capital Accord (also known as Basel II).

7. Internationalization is driving more complex internal and external corporate interdependencies. Boards should be concerned about the risk implications of new business partnering models, more diverse product and service portfolios, and international markets and operations. Interdependencies arising from these (and other issues) can radically alter the risk profile of a firm.

8. See, for example, Whittaker (1999). Also many of the research reports by the Standish Group (www.standishgroup.com) discuss this issue.

9. See Browning (1993).

10. A 2003 survey by the global information technology industry research firm, Gartner found that in 2003, large financial services firms intended to spend in the range of $500,000 to $2.5 million on information technology specifically aimed at risk management. This, according to the survey, was around 9.2% of the average 2003 information technology (IT) budget. See Gartner (http://www.gartner.com/press_releases/pr22apr2003a.html).

11. Jean de La Fontaine, born around 1621, is the most celebrated French poet. The passage is taken from his fable "The Mandrake."

12. An enterprise system (ES) is a corporate-wide vendor-supplied information sys-tem. It integrates all financial figures, accounting information, human resource information, supply chain information, and customer information into a central database. Like vendor-supplied risk management systems, an ES is essentially an information technology application that serves key corporate functions and involves centralized information shared by many users.

13. Typically, a business case contrasts the limitations of current risk identification, measurement, analysis, management, reporting, control, and regulatory requirements against the perceived benefits of a vendor-supplied system.

14. See, for example, Katz (1982a), Vaill (1982), Keller (1986), Lamb (1985), Larson and LaFasto (1989), Keller (1992), Browning (1993), Whittaker (1999), Amabile et al. (2001), and Dove (2004).

15. See, for example, Buckhout, Frey, and Nemec (1999), Bingi, Sharma, and Godla (1999), Rosario (2000), Willcocks and Sykes (2000), and Somers and Nelson (2001).

16. Two very practical Web sites dealing with the creation and characteristics of high performance teams are: (1) High Performance Teams (http://ptcpartners.com/Team/home.htm), a resource for businesses and organizations interested in harnessing the power of teams, and (2) The CEO Refresher (http://www.refresher.com/archives7.html), part of the CFO refresher Web site.

17. See, for example, Carmel and Bird (1997). Social network models can help explain how and why small teams work. See, for example, Buys and Larson (1979). And for a quantitative approach to the analysis of social networks, see Wasserman and Faust (1994).

18. See Thelen (1949).

19. See Campion, Medkger, and Higgs (1993) and Ziller (1957).

20. See Harkins and Petty (1982) and Karau and Williams (1993).

21. See, for example, Bingi, Sharma, and Godla (1999), Buckhout, Frey, and Nemec (1999), or Nah and Lau (2001).

22. A common question of those involved in vendor assessment and selection for the first time is where can we get an illustrative copy of an RFP. In the banking and financial services sector, intellectual property and nondisclosure clauses alongside other terms and conditions prohibit vendors from making RFPs publicly available. Fortunately, there are a number of sources in the public sector from which real and fully scoped RFPs can be viewed. The National Institutes of Health request for proposal directory has been established to provide Internet users with quick and easy access to RFP solicitations available at the National Institutes of Health. It provides direct links to all currently active electronic Request for Proposals issued by the National Institutes of Health. It is available at National Institutes of Health (http://ocm.od.nih.gov/contracts/rfps/MAINPAGE.HTM).

    The large North American cities are also a good source of illustrative RFPs. For example, in New York City, the Office of Labor Relations which represents the mayor in the conduct of all labor relations between the City of New York and labor organizations representing employees of the city, has a dedicated RFP site located at NYC Office of Labor Relations (http://www.nyc.gov/html/olr/html/requests/rfp.shtml).

    Two well structured textbooks which provide immense detail on the construction of robust RFPs are Porter-Roth and Young (2001) and Fria (2005).

23. Hugh Woodward, Chairman, Board of Directors, Project Management Institute (PMI). See PMForum (www.pmforum.org), a Web site dedicated to the exchange of project management information and knowledge.

24. See, for example, Brown and Adams (2000).

25. See, for example, Ibbs and Kwak (1997a, 1997b, 2000). In fact, to be more precise, these studies show a positive correlation between the level of project management maturity in an organization and what Ibbs and Kwak call "Project Management Return on Investment."

26. See, for example, Vaill (1982) and Amabile et al. (2001).

27. This chapter did not discuss in any detail issues surrounding the developing of an in-house risk management system. An interesting discussion as to why this is not necessarily a good idea can be found in two articles by Tom Groenfeldt; the first entitled, Why In-House Systems Fail and the second, Why Vendors Have An Edge. Both are freely available at Derivative Strategy (http://www.derivativesstrategy.com/).

10

Snake Oil Salesmen, Goat Gonads, and Value at Risk

We began this text by stating risk is a permanent loss of capital. Without a shadow of a doubt, we are absolutely convinced this holds true in all places and at all times. Yes, there are many other forms of financial risk, but first and foremost, investors and risk managers must be concerned about permanent loss. George Samuel Clason, the soldier, businessman, and author in his classic work, *The Richest Man in Babylon*, captured this essential truth.

> The first sound principle of investment is security for thy principal. Is it wise to be intrigued by larger earnings when thy principal may be lost? I say not. The penalty of risk is probable loss. Study carefully, before parting with thy treasure, each assurance that it may be safely reclaimed. Be not misled by thine own romantic desires to make wealth rapidly.

Eighty or so years after the publication of Clason's book, a text by Howard Marks entitled, *The Most Important Thing*, elucidated this important principle for a new generation of investors:

> Rather than volatility, I think people decline to make investments primarily because they're worried about a loss of capital or an unacceptable low return. To me, "I need more upside potential because I'm afraid I could lose money," makes an awful lot more sense than, "I need more upside potential because I'm afraid the price may fluctuate." No, I'm sure "risk" is—first and foremost—the likelihood of losing money.

Howard Marks is no ordinary author. He cofounded the investment house Oaktree Capital Management, which today manages over $80 billion, primarily in high yield and convertible bonds.

Investors in tulip bulbs during the Netherlands golden age may have wished the siren sounded by Mackay were louder: "Many a representative of a noble line saw the fortunes of his house ruined beyond redemption." Individuals reduced to desperate acts of beggary by the gluttonous thievery of Bernie Madoff are the inevitable realization of the violation of Clason's first principle. Risk managers, academicians, and investors should take careful note.

When risk, as we define it, manifests, and permanent loss occurs, it is too late. For this reason, in part, risk managers, quantitative authorities, and academics have devised ingenious ways to "measure" risk ahead of real-ized loss. There is no single concept which has captured the imagination of risk managers in this regard as Value at Risk (VaR). A Google search deliv-ered 1.6 million hits, BING retrieved close to 600 hundred thousand, and a scan of academic journals using Google Scholar delivered over 42 thousand research articles on the subject. This author himself is the researcher behind at least two of these![1] Today VaR dominates the risk management landscape. It is for this reason we devote an entire chapter to it. In doing so we offer a concise explanation of it, explore where it came from, and reveal the truths about it nobody dares tell you.

VaR Explained

VaR is founded on a simple concept. However, if you glance through any textbook on the subject you will invariably have to wade through numer-ous diagrams, equations, and the occasional mathematical lemma or proof. Fortunately, none of this is necessary to comprehend and appreciate the essence of the concept. Managers, boards of directors, trustees, and inves-tors can understand all they need regarding this concept by carefully read-ing this section. No diagrams, no equations, and certainly no mathematical lemmas or proofs. Instead, we focus on the easily digestible quintessence of the concept.

While there are many sources of risk facing a financial institution, profits are generally tied directly or indirectly to the behavior of economic growth, interest rates, and prices. The natural fluctuations in asset prices as a conse-quence of uncertainty surrounding these economic forces is known as mar-ket risk. Over a decade ago following a series of huge billion dollar plus losses in both the public and private sectors,[2] the question many were asking was how do we convey the degree of market risk inherent in our portfolios in a simple and straightforward way to senior management, investors, and regulators. The issue was explored by Thomas Linsmeier and Neil Pearson, professors at the University of Illinois at Urbana–Champaign[3]:

> You are responsible for managing your company's foreign exchange positions. Your boss, or your boss's boss, has been reading about deriva-tive losses suffered by other companies, and wants to know if the same thing could happen to his company. That is, he wants to know just how much market risk the company is taking. What do you say? You could start by listing and describing the company's positions, but this isn't likely to be helpful unless there are only a handful. ... Or you could talk about the portfolio's sensitivities, i.e. how much the value of the portfolio changes when various underlying market rates or prices change, and perhaps option delta's and gamma's. However, you are unlikely to win

favor with your superiors by putting them to sleep. ... You could simply assure your superiors that you never speculate but rather use derivatives only to hedge, but they understand that this statement is vacuous. They know that the word "hedge" is so ill-defined and flexible that virtually any transaction can be characterized as a hedge. So what do you say? Perhaps the best answer starts: "The value at risk is ..."

Professors Linsmeier and Pearson suggested VaR as the optimal solution because it addresses market risk by providing an estimate of loss with an associated probability. Indeed, VaR is often defined as *the loss in market value that may be sustained from an adverse movement in market prices over a specific time horizon and with a given degree of confidence.* The time horizon is often referred to as the holding period. Advocates of VaR argue it answers the question: *How much could the value of the portfolio decline over the next period of time?*

The choice of the holding period depends on the liquidity of the assets in the portfolio and how frequently they are traded. An investor in real estate will likely have a very different holding period than an options day trader. For the real estate investor, the holding period can be many years, for the day trader a matter of minutes. The holding period can be of any length, but it is assumed the portfolio composition does not change during the holding period. For large investors with liquid portfolios it is typically set to 1 day, 10 days, or 1 month. The *confidence level* determines the probability of loss, usually calculated at the 95% or 99% levels.[4]

Roughly speaking, VaR yields an estimate of the largest losses a portfolio is likely to experience in all but very exceptional trading days. As a simple illustration, consider the VaR estimate for a firm that holds a diverse portfolio of financial assets. At the end of the trading day, the market value of the portfolio can be determined; let us say £10 million. Suppose the firm reports that its portfolio has a 1 day VaR of £116,317 at the 99% confidence level. What does this number tell us?

A 99% degree of confidence implies approximately once in every 100 days the portfolio can expect to suffer a mark to market loss of at least £116,317. Notice that we say *at least* £116,317, the actual loss experienced could be a little larger, say £130,500, or much larger, say £350,500. If the firm had chosen a 95% level of confidence, the VaR estimate would have been smaller say £82,243. So in this case, approximately 5 trading days in every 100 we can expect to see a daily loss in market value of £82,243 or more. This single metric is appealing to the risk manager as it allows him to puff out his chest, look you in the eye, and declare with absolute confidence:

We are 99% certain that we will not lose more than £116,317 over the next trading day.

It is the reinterpretation of risk as the potential minimum loss that can occur over a certain time horizon with a specific probability that lies at the heart of VaR.

The Joyous Exclamation of Simons

It is, perhaps, a little easier to grapple with the issues surrounding VaR if one keeps in mind the astounding story of Dr. John Brinkley[5]:

> The remarkable events I'm going to chronicle here would likely never have unfolded, in 1917, if young Dr. John Brinkley had not been hired as house doctor at the Swift meatpacking company, located in Kansas. He was dazzled by the vigorous mating activities of the goats destined for the slaughterhouse. A couple of years later, after Brinkley had gone into private practice in Milford, Kansas, a farmer named Stittsworth came to see him. Stittsworth complained of a sagging libido. Recalling the goats' frantic antics, the doctor semi-jokingly told his patient that what he needed was some goat glands. Stittsworth quickly responded, "So, Doc, put 'em in. Transplant 'em." … Most doctors would have ignored the bizarre request, but Brinkley was not like most doctors. In fact, he wasn't a doctor at all. … He called himself a doctor on the basis of a $500 diploma he had purchased from the Eclectic Medical University of Kansas City, Missouri. … Brinkley went to work, implanting a bit of goat gonad in Stittsworth's testicle. Within weeks the farmer was back to thank the doctor for giving him back his libido … his [the Farmer's] wife gave birth to a boy, whom they appropriately named Billy. … Soon Brinkley's business was booming. … Brinkley was charging $750 per transplant, and he couldn't keep up with the demand. All men needed the Brinkley operation, he declared, but the procedure was most suited to the intelligent and least suited to the "stupid type." This, of course, ensured that few of his patients would admit that they had not benefited from the operation.[6]

Today, one hopes the discerning reader might be somewhat skeptical of Brinkley's claims. Alas, as we saw in the very first chapter, unreason exists in places where it should not! Back in 1917, neither scientific argument nor lack of medical credentials hindered Brinkley from capturing the public's attention. That he had a long narrow face with dark darting oblong eyes and fashioned a grizzly gray goatee beard only heighted the fascination. Discussions of Brinkley's xenotransplantation procedure were heard over the airwaves of crackly AM radio, written about in popular magazines, and spread by word of mouth like a contagious skin rash. At a whopping $750 per operation, Brinkley had firmly imprinted the value of his operation deep into the imagination of the American people; and this despite the utter and total lack of scientific evidence supporting the efficacy of his procedure.

To the great chagrin of the American Medical Association, the unquestioning, uncritical, and well-heeled rushed to Brinkley's door. Dollar bills could not be exchanged quickly enough in return for gonads. One imagines Brinkley was the character of person not overly concerned about the form of payment. Settling up in gold, silver, diamonds, and other hard assets, one feels, would have been readily acceptable to him. Brinkley became very wealthy indeed.

The stampede toward VaR is somewhat reminiscent of the crowds who flocked to Brinkley's surgery. Despite the monotonous bleating of the antagonists,[7] within little more than a decade, it has come to dominate the risk management landscape. Its supremacy epitomized in the joyous exclamation of Federal Reserve Bank of Boston economist Katerina Simons[8]:

> In many financial circles, the reputation of value at risk stands as high as that of motherhood and apple pie.

Well over a decade since Simons's statement, VaR remains high on the agenda of risk managers, board of directors, and regulators. It is now being used in a whole host of activities including risk reporting, risk limit setting, the calculation of regulatory capital, performance measurement, internal capital allocation, and asset allocation. Today the cry for VaR can be heard loud and clear, as the snake oil salesman Brinkley might say, "All corporations need VaR, but it is least suited to the stupid type."

The Tipping Point

Malcolm Gladwell, in his bestselling book *The Tipping Point*,[9] discusses the causes of epidemics. At the heart of his great treatise are three simple concepts:

> Epidemics are a function of the people who transmit infectious agents, the infectious agent itself, and the environment in which the infectious agent is operating. And when an epidemic tips, when it is jolted out of equilibrium, it tips because something has happened, some change has occurred in one (or two or three) of those areas. These three agents of change I call the Law of the Few, the Stickiness Factor, and the Power of Context.

The Law of the Few refers to the observation that a very few people are responsible for causing ideas to spread and take hold. These individuals, referred to by Gladwell as Salespeople, Mavens, and Connectors, are the influencers in society. As must have happened with Farmer Stittsworth's good news of his miraculous child Billy, VaR captured the imagination of Connectors, Mavens, and Salespeople.

In 1990, VaR was novel. Few financial institutions utilized the metric. By the mid 2000s, its global dominance was albeit complete; scholars had traced out in sedulous detail its historical origin[10]; disciplines which had nothing whatsoever to do with finance scurried to find an application for it in their field[11]; and risk managers across the globe seemed completely enthralled by it. Wave after wave of risk analysts have been certified in the intricate details of VaR. Quantitative analysts old and young can make a handsome living creating new ways to model it. Entire financial enterprises have been launched on the back of it. Quixotic academics spend years creating new ways to theorize about it, regulators demand its calculation and boards of directors insist on it being reported. The rise has been so meteoric, its proponents so energetic, many younger risk analysts schooled principally in the modern tools of the discipline take it as a given.

VaR as we know it today originated on the derivatives trading desks of investment banks, at the time, exotic areas little understood by the general population. As leverage became a key tool in return generation, trading firms sought new ways to manage risk taking. This motivated new metrics of risk. It seemed with VaR the science of risk measurement had reached new heights. Risk could be corralled as easily as skilled ranchers round up a herd of Texas Long Horned cattle. Once cornered, it could be systematically exploited to feed the appetite for outsized returns.

Seemingly without any serious challenge, protected by its own band of zealots, VaR spread rapidly from trading firm to trading firm and then out into the wider financial services community.[12] Typical is the headline, "Suncorp achieves Australian first for investment risk management." *Australian Banking & Finance* magazine reported:

> Suncorp Investment management has raised the bar in the measurement and reporting of investment market risk after becoming the first Australian fund manager to go live with a DST International (DSTi) risk management solution, HiRisk, that integrates value at risk into daily investment processes. ... The system sets and monitors value at risk limits on a daily basis to ensure Suncorp trades within its risk tolerance levels and provide patterns or indicators of risk for individual portfolios.[13]

Antagonists of VaR could only watch impotent, with growing chagrin, as what they perceived as something akin to Brinkley's snake oil gained ground to become the de facto industry risk metric. These purveyors of doubt complained bitterly, but nobody took any notice[14]:

> Critics of VaR (including the author) argue that simplification could result in such distortions as to nullify the value of the measurement. Furthermore, it can lead to charlatanism: Lulling an innocent investor or business manager into a false sense of security could be a serious breach of faith. ... The most nefarious effect of VaR is that it has allowed people who have never had any exposure to market risks to express their opinion on the matter.

Little attention was paid to the antagonists because VaR is sticky. Gladwell explains:

> Stickiness means that a message makes an impact. You can't get it out of your head. It sticks in your memory.

VaR is sticky, in part, because it attempts to directly answer the question, "How much market risk are we taking?" For managers, investors, and regulators, the answer offers a measure of how bad things can get. It is sticky, in part, because it can be estimated for any type of portfolio. For example, a proprietary trading unit might have portfolios of crude oil derivatives, interest rate swaps, currencies, and corporate bonds. VaR can be calculated on each of these separate portfolios and aggregated into a single number. It is sticky, in part, because it aggregates all of the risks in a portfolio into a single number, which can be easily conveyed to senior managers, directors, and regulators and disclosed in an annual report. VaR is a very sticky concept indeed.

Stickiness has helped it become a standard measure of risk, not only for financial institutions involved in large-scale trading, but also for retail banks, insurance companies, institutional investors and increasingly in non-financial enterprises. The National Commercial Bank's Board of Directors' Report captures this trend[15]:

> The Market Risk Management unit monitors on a live basis the risk taking activities of all the mark-to-market traders in the treasury. Each trader's position risk is measured using value at risk (the latest risk management technique). ... The risk of the trader (as well as the business unit) is compared against a pre-set daily limit that is decided during the preparation of the year's budget. The Market Risk Management unit also monitors and reports on a daily basis the profit and loss of each trader and each business unit and compares it against a maximum loss limit set at the start of the budget year. The Market Risk unit calculates the daily value at risk at a 2 standard deviation changes of prices (which means that there is only a 2.5% probability that the daily outcome will be worse than the pre-set limits).

The context of risk management shifted profoundly in the late 1990s and early 2000s. Alongside the rise of VaR a new breed of professional financial risk management organizations, with a bewildering array of certifications, began to emerge.[16] Emanating principally from the financial engineering community and founded on the principles of quantitative finance, their members have swiftly come to dominate the risk management debate. At the same time those who traditionally viewed themselves as overseers of risk such as actuaries and accountants struggled to keep pace with this new breed of überquantitative professional.[17]

The changing financial regulatory landscape also provided impetus for the expansion of VaR. In 1993, the Bank of International Settlements members met in Basle. They amended the so-called Basle Accord to require banks and some other financial institutions to hold in reserve capital to cover 10 days of potential losses. A 10-day 95% VaR framework served as the foundation for the reserve capital calculation. More recently, in the European Union (EU) under Article 21 of the Undertaking in Collective Investments and

Transferable Securities[18] (commonly known as UCITS III), certain investment funds (known as sophisticated UCITS) were required to[19]:

> ... employ a risk-management process which enables it to monitor and measure at any time the risk of the positions and their contribution to the overall risk profile of the portfolio ...

The precise meaning of a *risk management process* and the role of VaR within such a process is elucidated in the European Commission Recommendation 2004/383/EC where it was suggested[20]:

> In the case of "sophisticated UCITS," Member States are recommended to require management or investment companies to apply regularly VaR approaches. In the VaR-approaches, the maximum potential loss that a UCITS portfolio could suffer within a certain time horizon and a certain degree of confidence is estimated. ... For the application of VaR-approaches, Member States are recommended to require the use of appropriate standards in conformity with point 3.1. For this purpose, Member States should consider, as a possible reference the following parameters: a 99% confidence interval, a holding period of one month and "recent" volatilities, i.e., no more than one year from the calculation date without prejudice to further testing by the competent authorities.

What the *Rocket Scientists* May Not Tell You, But You Need to Know

For many senior managers, board members, and investors, VaR remains somewhat of an enigma—*the latest risk management technique* calculated by teams of quantitative technocrats whose first language is mathematics rather than English. Little digestible knowledge (to nonquantitative individuals) is offered from textbooks or academic papers, for they too are stuffed full of statistical terminology, mathematical equations, lemmas, and conjectures. The scientific sounding jargon does little to lift the erudite haze that surrounds the subject. The unfortunate reality is that much of the debate about the critical issues surrounding VaR is simply inaccessible to many interested parties who must make use of it.

Limited knowledge about critical aspects of your risk can, as Brinkley would attest, prove very troublesome. One of the first signs of trouble ahead for him came when he decided to use Angora goat testicles instead of those from his usual Toggenburg goats. Unfortunately, men who received the said testicles found themselves singly unable to exercise their "wondrous increase in" libido—no woman could bear to be within 50 feet of them—for as Brinkley himself lamented: "They reeked like a steamy barn in midsummer."[23]

As any capriculturist will tell you, the differences between an Angora and Toggenburg are quite profound. The Toggenburg is a sturdy vigorous dairy goat originally from the Toggenburg Valley in Switzerland. The Angora goat, on the other hand, is prized for its lustrous long mohair and little else. Brinkley's choice of Angora is, on the face of it, a little puzzling because they are substantially less prolific in their mating activities than the Toggenburg. The point being an individual who is well informed about goats would tend to prefer the Toggenburg over the Angora on issues of libido. Brinkley, it seems, was not well informed about goats.

Fortunately, the consequence of Brinkley's limited knowledge was little more than a rather pungent odor for a number of his well-heeled patients. The consequence of lack of knowledge by decision makers and senior management about the inherent characteristics and delimit of their VaR model is on an altogether different scale. The 1998 failure and U.S. $4.6 billion losses of the hedge fund Long Term Capital Management[21] have been attributed by some authors to a poor understanding of the limits of their VaR model.[22] Senior management, executives, and board members should have intimate knowledge of their value at risk.

Unfortunately, the perceived complexity of VaR, unfamiliarity with the underlying statistical concepts, and uncertainty about where to start, may hinder boards, managers, and trustees from seeking answers to important questions. Only when the *rocket scientist's* model "begin to smoke" and turns uncontrollably upon its creator, somewhat like Mary Shelley's Frankenstein monster, do the penetrating questions arise. And this, alas, may be too late. There are significant issues rocket scientists may be reluctant to disclose but you need to be aware of.

The Curse of the Bell-Shaped Curve

The calculation of VaR requires a number of inputs, which include historical data on market prices and rates, the current portfolio positions, and models for pricing those positions. These inputs are then combined in various ways depending on the method used to derive an estimate of VaR. As one might expect, the estimate will depend partly on the portfolio return and volatility. It will also depend on the probability distribution of portfolio returns, holding period, and level of liquidity of the underlying instruments or assets in the portfolio. Since we may not know the exact probability distribution of portfolio returns, it is common practice to select a known mathematical probability distribution as a proxy for the actual distribution. A popular choice is the bell-shaped curve or normal probability distribution.

Unfortunately, much of the existing literature has shown the distributions of numerous financial asset returns exhibit systematic deviations away from the bell-shaped curve.[25] Federal Reserve Board Chairman Alan Greenspan identified this as a key issue[26]:

THE DISCOVERY OF THE NORMAL DISTRIBUTION

The normal distribution was discovered by the Huguenot refugee, Abraham de Moivre, in around 1733; however it was Gauss (1809) in his *Theoria motus corporum* who derived it.[23] It rapidly became the most important probability distribution in the statistician's toolbox. The extraordinary Victorian polymath, Sir Francis Galton, who called it the "law of frequency of error," wrote of it:

> I know scarcely anything so apt to impress the imagination as the wonderful form of the cosmic order expressed by the "Law of Frequency of Error." The law would have been personified by the Greeks and deified if they had known of it. It reigns with serenity and in complete self-effacement amidst the wildest confusion. The huger the mob and the greater the apparent anarchy, the more perfect is its sway. It is the supreme law of Unreason.[24]

> ... as you well know, the biggest problem we now have with the whole evolution of risk is the fat-tail problem, which is really creating very large conceptual difficulties. Because, as we all know, the assumption of normality enables us to drop off the huge amount of complexity in our equations. ... Because once you start putting in non-normality assumptions, which is unfortunately what characterizes the real world, then these issues become extremely difficult.

Fat tails imply extreme losses occur much more frequently than predicted by the normal distribution, and as a result VaR models built using this distribution may underestimate market risk. In addition, many asset returns tend to be skewed to the left so that large negative returns are more likely than large positive returns. This violates the assumption of symmetry in asset returns embedded in the bell-shaped curve.

Deviations away from the bell-shaped curve pose a very challenging statistical problem. There has arisen a multitude of approaches that attempt to address this issue.[27] The commonality between the approaches is that they all follow a general structure: first, mark to market the portfolio, second, estimate the distribution of portfolio returns, and third, compute the VaR of the portfolio. Despite this, the issue is far from resolved.

Exact Imprecision—On the Accuracy of VaR

Consider the trading books of large banks, which contain tens of thousands of positions. To obtain an estimate of VaR requires some simplifying assumptions, as Jeremy Berkowitz, professor of finance at the University of Houston, points out[28]:

> To estimate the portfolio's risk structure, the banks make many approximations, and parameters are often estimated only roughly. While this may appear to give representation to a wide range of potential risks, the various compromises tend to reduce any forecasting advantage.

Of course, this raises the question of the accuracy of the VaR estimate. Yet, among the legion of quantitative technocrats hired to compute it, this issue is rarely if ever discussed—their job is to gather together the relevant data and produce a *corporate-wide* VaR estimate, and that is exactly what they do. This presents a missed opportunity. VaR is a statistical model. All statistical models need constant evaluation and testing to access their accuracy. This is true regardless of what the statistical model is used to measure or predict.

Berkowitz investigated the extent of this issue by using data gathered from six large bank holding companies. Many financial institutions develop their own in-house VaR model. His study was the first to provide direct evidence on the performance of such models for large trading firms. The results were surprising and a little disturbing. Despite the considerable information collected by the banks during the process of deriving their estimate of risk, Berkowitz finds "...the reported VaRs are less useful as a measure of actual portfolio risk." The metric derived to address concerns over a permanent loss of capital, the metric tailored to address questions surrounding the probability of loss, in practice, according to this study has residual utility as an actual measure of portfolio risk! This is an astounding finding.

A decade has passed since Berkowitz's observation. His results are well known among rocket scientists and risk managers, yet are rarely discussed. Part of the explanation may possibly be found in the fable of Spreadsheet City discussed in Chapter 8. There, we observed the software engineers were simply having too much fun programming and reprogramming the software to worry about its ultimate use.

Berkowitz in the same research article discovers a very simple statistical model without the *bangs and flashes* typically favored by rocket scientists, which provides a more accurate measure of tail risk than the large scale VaR models used by the banks.[29] Perhaps this serves to underscore the fact that there has yet to be articulated a unique, universal, and widely accepted basis for constructing VaR. There is yet to emerge, in the VaR literature, any degree of hegemony. The battles between the various schools of thought continue[30]; the victors have yet to be declared. The battleground is a complicated place because it inherits from the academics the idealized intuitive notion of VaR, which if only it can be properly constructed, will provide an effective tool for the management of market, credit, and indeed all other risks. Unfortunately, the optimal method for implementing the concept remains far from settled.

This raises a particularly important issue. Even though a VaR model is grounded in statistical and mathematical principles, it is also to a significant extent influenced by subjective opinions and unavoidable approximations.

The VaR model builder must make judgments about the key risk factors, their distributional behavior, and the observation periods over which they are relevant. Yet, few if any practitioners openly acknowledge this or document or make known the consequence of their assumptions. Oftentimes, risk managers themselves are unaware of the importance of the issue. Those who are more prescient may elect not to make their superiors aware of this issue. Yet, such knowledge could provide valuable insight into the functioning of their model, in particular, its sensitivity, robustness, and quality. Without such explicit detail, one may well be left, like Saint Augustine, wondering:

> For so it is, O Lord my God, I measure it! But what it is I measure, I do not know.

Risk managers need to constantly evaluate their VaR model. The old adage holds as true for risk management as it does for any other area of business—measure what you want, but reward what you measure. Risk managers should prepare regular reports on the efficacy of their VaR model. It is important to understand that all VaR models are not equal.[31]

> ... we argue that institutions are too dependent on one single VaR estimate. A more critical review is needed. Given the high reliance on VaR estimates, evaluating the accuracy of the underlying VaR models is a necessary exercise. Further, it is important to test how different assumptions affect the VaR forecast, and then evaluate whether some assumptions are more suitable for certain kinds of portfolios than for others.

The statistical tools are now widely available to do this. The "Annual Assessment of Our VaR Model" report should be prepared, submitted, and defended by the risk management team. There are two significant benefits of doing this. First, senior management may lack confidence in the output from the model unless its efficacy is systematically documented. Second, in the spirit of continuous process improvement, it will spur the creation of more accurate, robust, and value-added risk modeling.

In the end, it is important to realize your VaR model is likely to be inaccurate, backward-looking, and dependent on a wide range of possibly unknown (to you) qualitative assumptions and personal biases. It will not save you when risk strikes. Nor will any other risk metric. For risk is a permanent loss of capital and this is a fundamental rule of risk management.

For Further Thought

A number of issues are worthy of additional discussion:

- Do your investment professionals use VaR in their risk taking? The answer will tell you much about the utility of this metric.
- How well do your current VaR models capture the behavior of the tails of the distribution of profit and loss?
- If something goes really wrong, how much money are you likely to lose?
- How does your risk group go about assessing the probability that large losses will occur and the extent of losses in the event of unfortunate movements in markets?
- How does your risk group assess the accuracy and performance of its VaR model?

Additional Resources

For an elementary introduction to VaR, see Simons (1996) or Jorion (2000). Further discussion of various VaR methods can be found in Duffie and Pan (1997), Venkataraman (1997), Boudoukh, Richardson, and Whitelaw (1998), Huisman, Koedijk, and Pownall (1999), Johansson, Seiler, and Tjarnberg (1999), Abken (2000), Billio and Pelizzon (2000), Fan and Gu (2003), Albanese, Jackson, and Wiberg (2004), Ming-Yuan and Hsiou-Wei (2004), Gilli and Këllezi (2006), or Pritsker (2006). See Glasserman (2004), Glasserman and Li (2003), or Antonelli and Iovino (2002) for discussion of advanced numerical methods and implementation. Feridun (2005) outlines lessons for VaR from the failure of the hedge fund Long Term Capital Management. Additional historical context can be found in Hartmann (1996) and Holton (2002). Berkowitz and O'Brien (2002) discuss the accuracy of large-scale corporate VaR models. The use of VaR outside of the financial service industry is illustrated in Koch (2006). Further discussion of the nature of asset price returns can be found in Fama (1965), Gray and French (1990), or Bekaert, Erb, Harvey, and Viskanta (1998). Also, see the classical work of Galton (1889). Econometric approaches to model asset price and portfolio volatility are outlined in the classic papers of Engle (1982) and Bollerslev (1986). De Marchi and Gilbert (1989) discuss the relationship between methodology and practice. Further details on the extraordinary life of John Brinkley can be found in the fascinating book by Lee (2002).

Abken, P. (2000). An Empirical Evaluation of Value-at-Risk by Scenario Simulation. *Journal of Derivatives* 7(4):12–29.

Albanese, C., Jackson, K., and Wiberg, P. (2004). A New Fourier Transform Algorithm for Value-at-Risk. *Quantitative Finance* 4(3) (June):328–338.

Antonelli, S. and Iovino, M. (2002). Optimization of Monte Carlo Procedures for Value at Risk Estimates. *Economic Notes* 31(1):59–78 (20).

Bekaert, G., Erb, C., Harvey, C., and Viskanta, T. (1998). Distributional Characteristics of Emerging Market Returns and Asset Allocation. *Journal of Portfolio Management* 24:102–15.

Berkowitz, J. and O'Brien, J. (2002). How Accurate Are Value-at-Risk Models at Commercial Banks? *Journal of Finance* 57(3):1093–1111.

Billio, M., and Pelizzon, L. (2000). Value-at-Risk: A Multivariate Switching Regime Approach. *Journal of Empirical Finance* 7:531–554.

Bollerslev, T. (1986) Generalised Autoregressive Conditional Heteroskedasticy. *Journal of Econometrics* 31:307–327.

Boudoukh, J., Richardson, M., and. Whitelaw, R.F. (1998). The Best of Both Worlds: A Hybrid Approach to Calculating Value at Risk. *Risk* 11.

De Marchi, N. and Gilbert, C. (1989). *History and Methodology of Econometrics*. Oxford: Oxford University Press.

Duffie, D. and Pan, J. (1997). An Overview of Value at Risk. *Journal of Derivatives* 4:7–49.

Edwards, F.R. and Canter, M.S. (1995). The Collapse of Metallgesellschaft: Unhedgeable Risks, Poor Hedging Strategy, or Just Bad Luck? *Journal of Applied Corporate Finance*, Spring.

Engle, R.F. (1982). Autoregressive Conditional Heteroscedasticity with Estimates of the Variance of UK Inflation. *Econometrica* 50:987–1007.

Fama, E.F. (1965). The Behavior of Stock Market Prices. *Journal of Business* 38:34–105.

Fan, J. and Gu, J. (2003). Semiparametric Estimation of Value at Risk. *Econometrics Journal* 6:261–290.

Feridun, M. (2005). Value at Risk: Any Lessons from the Crash of Long-Term Capital Management (LTCM)? *Journal of Business Administration Online* Spring 4(1).

Galton, F. (1889). *Natural Inheritance*. London and New York: Macmillan and Company.

Gilli, M. and Këllezi, E. (2006). An Application of Extreme Value Theory for Measuring Financial Risk. *Journal Computational Economics* 27(2–3) (May):207–228.

Gladwell, M. (2002). *The Tipping Point. How Little Things Can Make a Big Difference*. New York: Back Bay Books.

Glasserman, P. and Li, J. (2003). Importance Sampling for a Mixed Poisson Model of Portfolio Credit Risk. In *Proceedings of the 2003 Winter Simulation Conference*, Chick, et al., (eds). Piscataway, NJ: IEEE Press.

Glasserman, P. (2004). Monte Carlo Methods in Financial Engineering, Number 53. In *Applications of Mathematics*. New York: Springer.

Gray, B. and French, D. (1990). Empirical Comparisons of Distributional Models for Stock Index Returns. *Journal of Business, Finance, and Accounting* 17:451–459.

Greenspan, A. (1997). Maintaining Financial Stability in a Global Economy. Discussion at the Federal Reserve Bank of Kansas City Symposium.

Hartmann, P. (1996). A Brief History of Value-at-Risk. *The Financial Regulator* 1(3):37–40.

Holton, G.A. (2002). History of Value-at-Risk: 1922–1998, Economics Working Paper Archive. Boston: EconWPA.

Huisman, R., Koedijk, C., and Pownall, R. (1998). VAR-x: Fat Tails in Financial Risk Management. *Journal of Risk*.

Johansson, F., Seiler, M.J., and Tjarnberg, M. (1999). Measuring Downside Portfolio Risk. *Journal of Portfolio Management* 96(1):26.1.

Jorion, P. (1995). *Big Bets Gone Bad: Derivatives and Bankruptcy in Orange County. The Largest Municipal Failure in U.S. History*. Bingley, England: Emerald Group Publishing.

Jorion, P. (2000). *Value-at-Risk: The New Benchmark for Managing Financial Risk*. McGraw-Hill.

Koch, S. (2006). Using Value-at-Risk for IS/IT Project and Portfolio Appraisal and Risk Management *Electronic Journal of Information Systems Evaluation* 9(1):1–6.

Lee, R.A. (2002). *The Bizarre Careers of John R. Brinkley*. Lexington: University Press of Kentucky.

Leeson, N. and Whitley, E. (1996). *Rogue Trader: How I Brought Down Barings Bank and Shook the Financial World*. London, England: Little Brown and Company.

Lewis, N.D., Okunev, J., and White, D. (2007). Using a Value at Risk Approach to Enhance Tactical Asset Allocation. *Journal of Investing* 16(4):15–19.

Lewis, N.D. and Okunev, J. (2009). Using Value at Risk to Enhance Asset Allocation in Life-Cycle Investment Funds. *Journal of Investing* 18(1):87–91.

Linsmeier, T. and Neil, P. (1996). *Risk Measurement: An Introduction to Value at Risk.* Working paper of the University of Illinois.

Ming-Yuan, L.L. and Hsiou-Wei, W.L. (2004). Estimating Value-at-Risk via Markov Switching ARCH Models—an Empirical Study on Stock Index Returns. *Applied Economics Letters* 11(11) (September 15):679–691.

Pritsker, M. (2006). The Hidden Dangers of Historical Simulation: Value-at-Risk Computation Methods in Portfolio Management. *Journal of Banking and Finance* 30(2):5.

Simons, K. (1996). Value at Risk—New Approaches To Risk Management. *New England Economic Review* (September/October).

Taleb, N. (1997). *Dynamic Hedging*. New York: John Wiley & Sons.

Venkataraman, S. (1997). Value at Risk for a Mixture of Normal Distributions: The Use of Quasi-Bayesian Estimation Techniques. *Economic Perspectives* 21.

Endnotes

1. See Lewis, Okunev, and White (2007) and Lewis and Okunev (2009).
2. Most spectacular were the Orange County failure, Barings Bank collapse, and Metallgesellschaft hedging miscalculation. See Jorion (1995), Leeson and Whitley (1996), and Edwards and Canter (1995).
3. See Linsmeier and Pearson (1996).
4. Note the probability of loss is equal to 1-confidence level. So, a confidence level of 99% is equivalent to a probability of loss equal to 1%.
5. For further details on the fascinating life and times of John Brinkley, see the entertaining book by Lee (2002).
6. Taken from Quackwatch (quackwatch.org).
7. VaR has a number of limitations. We discuss these later in the chapter.
8. Katerina noted by 1996, VaR had become an integral part of banking risk management. Regulators and practitioners appeared to have accepted it as the *right* way to measure risk. See Simons (1996).
9. See Gladwell (2002).
10. See, for example, Hartmann (1996) and also Holton (2002).
11. See, for example, Koch (2006).
12. Indeed, widespread interest in VaR as a risk management tool can be traced to a number of events in the early to mid-1990s: (1) The release to the general

public by JP Morgan of the full technical details of their VaR model (known as RiskMetrics™) during October 1994; (2) The Basle committee of Banking supervision reform of January 1996 which introduced VaR to measure market risk and used it to determine the regulatory capital charge. This regulatory capital was to be a cushion for banks on balance sheet and off balance sheet positions against unforeseen movements in market prices and interest rates; and (3) The European Union's Capital Adequacy Directive, which came into force in 1996 and allowed VaR models to be used to calculate the capital requirements for foreign exchange positions.

13. *Australian Banking and Finance* magazine, August 15, 2005.
14. See Taleb (1997).
15. For the year 2002. For further details, see National Commercial Bank, also known as Alahli Bank.
16. See the International Financial Risk Institute (www.ifri.ch), whose Web site has links to a number of professional risk management organizations.
17. For now at least, the actuaries and accountants' stranglehold on risk has been broken; and they are silent, scattered across the corporate landscape, tattered and torn like some once mighty, now defeated, army.
18. UCITS established a European *passport* for fund managers such that provided a fund is certified in one EU country, it may be marketed in the rest of the EU.
19. See Directive 2001/108/Ec of The European Parliament and of the Council of 21 January 2002 contained in the *Official Journal of the European Communities* (http://eur-lex.europa.eu/en/index.htm).
20. See Corrigendum to Commission Recommendation 2004/383/EC of 27 April 2004 on the use of financial derivative instruments for undertakings for collective investment in transferable securities (UCITS) in the *Official Journal of the European Union.*
21. By now the tale of Long-Term Capital Management (LTCM) is well known. A group of bond traders joined forces with Nobel Laureate academics to create a hedge fund with the intention of making lots of money. They failed spectacularly. The Federal Reserve Bank of New York had to facilitate a bailout of the LTCM, fearing liquidation might damage the global financial markets.
22. See, for example, Feridun (2005).
23. Mathematicians and physicists in his honor refer to it as the Gaussian distribution.
24. Galton (1899), page 66.
25. See, for example, Fama (1965), Gray and French (1990) or Bekaert, Erb, Harvey, and Viskanta (1998).
26. See Greenspan (1997).
27. For example, Johansson, Seiler, and Tjarnberg (1999) discuss 20 of the most common techniques. Albanese Jackson, and Wiberg (2004) use a Fourier transform method, Ming-Yuan and Hsiou-Wei (2004) propose a Markov Switching autoregressive conditional heteroskedasticity model, Venkataraman (1997) suggests the use of Quasi-Bayesian Estimation Techniques, Billio and Pelizzon (2000) use a multivariate switching regime volatility model, Fan and Gu (2003) turn to semiparametric estimation; since VaR is defined as a low quantile in the distribution of financial profits and losses, Gilli and Këllezi (2006), among others, explore the use of Extreme Value Theory. Boudoukh, Richardson, and Whitelaw (1998) discuss hybrid techniques.

28. See Berkowitz and O'Brien (2002).
29. Equally troubling was Berkkowitz and O'Brien's finding that VaR models failed to provide accurate forecasts of changes in profit and loss volatility. Indeed, the authors demonstrate that VaR forecasts based on the very parsimonious Generalized Autoregressive Conditional Heteroscedasticity (GARCH) models that were introduced by Engle (1982) and Bollerslev (1986) are better able to capture time-variability in profit and loss volatility. The authors show GARCH models produce lower VaR estimates and capture volatility clustering so that losses in excess of VaR were fewer in number and much less extreme.
30. For example, see Venkataraman (1997), Johansson, Seiler, and Tjarnberg (1999), Billio and Pelizzon (2000), Fan and Gu (2003), Albanese, Jackson, and Wiberg (2004), Ming-Yuan and Hsiou-Wei (2004), or Gilli and Këllezi (2006).
31. See Johansson, Seiler, and Tjarnberg (1999).

Index

A

ACCC; *See* Australian Competition and Consumer Commission
Ackerman, Gary, 48
Adequately capitalized institutions, 40
Ali, Muhammad, 49
Allied Irish Bank, 128
Alt-A loan market, 33
Alt-A mortgage, 37
Amaranth Advisors, 57, 62, 63, 112
American Dream, 125
Australian Competition and Consumer Commission (ACCC), 79
Australian Securities Exchange (ASX), 81
Availability cascade, 125

B

Bank of America, 10
Baptist Foundation of Arizona (BFA), 22, 23
BASEL II guidelines, 15
Basle Accord, 207
Bear Stearns, 10, 55, 128
Behavioral biases; *See* Strategic security director
Bell-shaped curve, 209–210
BFA; *See* Baptist Foundation of Arizona
Big Wing Theory, 38
Blue Dog Democrat, 47
Boardroom capitulation, 88
Brinkley, John, 204
British Mycological Society, 4
Buffet, Warren, 22; *See also* Warren Buffet principle of risk management
Bull-market rush, 10
Bush, George W., 84

C

California Public Employees' Retirement System (CalPERS), 87
CAMELS composite rating, 39
Capital Asset Pricing Model (CAPM), 58, 67–68
Carroll, Lewis, 17
Catastrophic events, 124
Cayley, Dorothy, 4
CDO; *See* Collateralized debt obligation
Chancery Lane, 128
Cheung, Meaghan, 44
Chief executive officer (CEO), 73, 141
Chief finance officer (CFO), 127
Chief investment officer (CIO), 53
Chief risk officer (CRO), 116, 131, 145
Citigroup, 113
CJD; *See* Creutzfeldt-Jakob disease
C++ language, 155
Clason, George Samuel, 201
Collateralized debt obligation (CDO), 75
Comply or explain approach, 81
Confidence level, 203
Conflicts of interest, 117
Connecticut Retirement Plans and Trust Funds, 86
Coolidge, Calvin, 19
Corporate governance, what textbooks will not tell you about, 73–109
  benefit of wolf pack capitalism, 86–88
    boardroom capitulation, 88
    Connecticut Retirement Plans and Trust Funds, 86
    financial transparency, 86
    hedge funds, 87
    proxy votes, 86
    shareholder activism, 86
  codes, 99
  collapse of house prices, 74

collateralized debt obligation, 75
cost of corporate governance, 93–95
  Greece, 94
  hostility to prescriptive
    legislation, 93
  non-prescriptive approach, 93
  policy makers, 94
credit markets, 74
essence of governance issue, 76–77
  confidence in capital markets, 76
  market mechanisms, 77
  shareholders, 77
fingers in public till, 76
foreclosure filings, 74
inherent ethos of risk management,
    88–92
  board of directors, responsibility
    of, 90
  business ethics, 90
  company failures, root of, 88
  compliance, 88
  potential threats, 89
  shareholder value, 92
  stakeholders, 89
  successful companies, 89
internal e-mails, 75
key points, 100
malfeasance, 73
Mighty Sparrow, 73, 76, 84
moat-gate scandal, 76
Ponzi scheme, 74
principles of corporate governance,
    91
questions, 101–102
retirement savings, 74
risk management, 92
robber barons, 73
rock star style executive excess, 73
role of criminal penalty, 83–85
  indecent spectacle, 83
  organic link, 84
  parade of catastrophe and
    disgrace, 84
  punishment, 85
  shareholders, 85
superficiality of compliance, 77–80
  chief compliance officer, 78
  definition of corporate
    governance, 78

disconnect, 79
  self-aggrandizement, 78
  swagger, 78, 80
value-added activities, 92
why "gentleman's" agreements do
    not work, 80–83
  best practice, 80, 82
  comply or explain approach, 81
  downsizing, 82
  forward thinking institutions, 80
  governance codes, 82
  Hampel Code, 80
  if not, why not approach, 81
  pension holidays, 82
  scandals, 83
  self-regulation, 80, 82
  voluntary adoption of corporate
    governance, 81
why governance failures are
    inevitable, 95–99
  audits of management
    performance, 95
  board of directors, 99
  common misconception, 97
  heavyweight legislation, 95
  lure of bigger money, 98
  proxy statement, 96
  scoundrels, 98
  self-deceit, 98
  temptation, 98
Cotton, Henry, 55
Countrywide, 74, 77
Creutzfeldt-Jakob disease (CJD), 124
CRO; *See* Chief risk officer

D

Daewoo, 83
Dahl, Roald, 177
Decompiler, 166
Deutsch, David, 60
Dickens, Charles, 9
Dochow, Darrel W., 31
Downsizing, 82

E

End-user computing, 157
Enron, 54, 62, 77, 85, 90, 93, 117

Ethology, 29
Ethos of risk management, 116
Euler's identity, 86
Eye watering failures, 180

F

Failure, understanding the nature of,
    177–199
  career risk, 177
  clarifying your requirements,
      185–189
    background information, 189
    Request for Proposal, 188
    Risk Information Requirements
        Table, 186
    Risk System Requirements
        Documentation, 186
    statement, 185
    threshold criteria, 189
    value at risk, 186
  critical role of executive buy-in,
      184–185
    empirical study, 185
    vendor choice, 184
  developing a winning game plan,
      181–182
    benchmark, 181
    expectations, 182
    plan, 181
  Gramm-Leach-Bliley Act of 1999, 197
  high performance team, creation of,
      182–183
    shared values, 183
    team direction, 183
    vendor selection, 182
  how to guarantee success by
      understanding the nature of
      failure, 179–181
    eye watering failures, 180
    loss of goodwill, 181
    vendor selection, 179
    zeitgeist, 180
  important lesson of $\frac{1}{2} \times n \times (n - 1)$,
      183–184
    communication link, 184
    social loafing, 184
  questions, 193–194
  risk systems vendor, 191–194
    truth about project managers,
        189–191
    value added of vendor risk
        information systems, 178–179
      business strategy, 178
      productivity and connectivity
          gains, 179
      seamless integration, 178
      technical hurdles, 179
    vendor selection, 178, 192
Fannie Mae; *See* Federal National
    Mortgage Association
Farnese, Alessandro, 3
Fastow, Andrew, 98
Fat-tail problem, 210
Fayol, Henry, 126; *See also* Strategic
    security director
Federal Deposit Insurance Corporation,
    34
Federal Deposit Insurance Corporation
    Improvement Act (FDICIA), 93
Federal Home Loan Bank System, 34
Federal National Mortgage Association
    (Fannie Mae), 116, 117, 128
Fidelity Magellan Fund, 160
Flippers, 16
Flynn Effect, 7–12
  average intelligence of the world, 8
  bull-market rush, 10
  gullibility, 8
  Intelligence Quotient, 8
  Madoff's victims, 11
  Ponzi scheme, 9
  signature crook, 9
  split-strike conversion strategy, 10
Ford Motor Company, 142, 145
Foreclosure filings, 74
Foreman, George, 49

G

Galton, Francis, 210
Gartner, 62
Glad Game, 12–17
  BASEL II guidelines, 15
  contempt of risk, 15
  delusions, 14
  flawed self-assessment, 15
  flippers, 16

incentivized speculation, 16
irony, 17
Johari window, 14
joke among mortgage brokers, 16
liar's loan, 16
mortgage backed securities, 15
natural disasters, 14
no money down mortgages, 16
Pollyanna, 12, 16
real estate "rags to riches" articles, 16
"regular Joe" investor, 16
subprime mortgage borrowers, 15
tax code, 16
unreason, 16
Yorkshire phrase, 16
Global Crossing, 90
Golden rule of risk management, 114, 115
Goldman Sachs, 10, 75, 77
Governance; *See* Corporate governance, what textbooks will not tell you about
Government assistance, 130
Gramm-Leach-Bliley Act, 197
Great Recession, 55
Greed; *See* Unreason, greed and
Greenspan, Alan, 209
Greenspan, Stephen, 8
Group of Thirty, 62

H

Hampel Code, 80
Hazard risks, 127
HealthSouth, 160
Hedge funds
bias, 124
competition, 58
corporate governance, 87, 103
fee structure, 87
large, 69
model, 118
multistrategy, 57
pressure, 88
sophisticated, 117
Held for sale portfolio, 38
Henninger, Daniel, 82
Hill, Napoleon, 98

HIV; *See* Human immunodeficiency virus
Holding period, 203
Holton, Glyn, 118
Homicide, 124
Hoover, Herbert, 22
Hot money, 38
Hudson, Mike, 35
Human immunodeficiency virus (HIV), 124
Hunter, Brian, 58
Hydro One, 145

I

IBM, 157
If not, why not approach, 81
Important lesson, 111–121
golden rule, consequences of ignoring, 115–117
business strategy, 115
ethos of risk management, 116
mortgage acquisitions, 117
risk appetite, 116
immutable condition for success in risk management, 117–119
conflicts of interest, 117
golden rule, 118
independence, 117
sophisticated hedge fund, 117
subprime mortgage crisis, 118
Odysseus and the Sirens' song, 114–115
golden rule of risk management, 114
guile, 115
off-balance-sheet partnership, 113
questions, 119–120
risk appetite, 111
risk assessment and control department, 112
sin eater, 111
sweetheart deal, 113
temptation, 113
unthinking risk managers, 111
value-added key points, 119
IndyMac, 33, 55, 63, 74
Ingham, Harry, 14

Integrated risk management; *See*
    Monocle on risk, advantage of
Intelligence Quotient (IQ), 8
Internet
    delivery of mortgage products, 34
    security, 127

J

Johari window, 14, 15
John Innes Horticultural Institution, 5
Jones, Robert, 58, 111

K

Keating, Charles, 42
Knee jerk cynicism, 25
Kreuger, Ivar, 17

L

Law of the Few, 205
Law of frequency of error, 210
Lay, Kenneth, 85, 98
Lehman Brothers, 10, 55, 63, 77, 128, 145
Levin, Carl, 75
Liar loans, 16, 50
Liberty bonds, 20
Lloyd's of London, 95, 144
Long-Term Capital Management
    (LTCM), 216
Lorenz, Konrad, 29, 49
Luft, Joseph, 14

M

Madoff, Bernard, 9, 24, 46, 74, 201
Maleficent hand of men in gray suits;
    See Regulators
Maounis, Nicholas M., 58
MARHedge multistrategy hedge fund
    Performance Awards, 61
Market risk, 59, 202
Marks, Howard, 201
Match King, 19
Merrill Lynch, 10
Milton, John, 57
Mitchell, Austin, 82
Moat-gate scandal, 76

Monocle on risk, advantage of, 139–154
    challenge, 144–146
        buffering of earnings, 145
        cultural roots of organization, 144
        lack of commitment, 145
    definition of monocle on risk, 140–141
        integration, 140
        new paradigm, 140
        risk-return decision, 141
    diffusion of risk management
        knowledge, 139
    essential elements of successful risk
        integration, 146–148
        competitive advantage, 148
        frameworks, 147
        guidelines, 146
        inevitable differences, 146
        risk tolerance, 147
    integrated risk management,
        148–151
    need for better risk management,
        143–144
        business as usual, 143
        coherent risk management, 143
        information, 144
    nightmare, 139
    questions, 150–151
    risk management silos, hidden
        dangers of, 141–143
        avoidable mistake, 142
        competitive advantage, 142
        decision to hedge, 141
        enemy action, 143
        inappropriate hedging, 142
    risk scent, 140
Monopoly rights, 21
Morgan Stanley, 10
Mortgage
    acquisitions, 117
    Alt-A, 37
    backed securities, 15
    brokers, joke among, 16
    IndyMac loans, 33
    no money down, 16
    products, Internet delivery of, 34
    refinancing of, 35
    subprime, 15, 34
    toxic, 75
Multistrategy hedge fund, 57

N

National Air Traffic Services (NATS), 190
Nobel Peace Prize, 22
No doc procedure, 36
No money down mortgages, 16
Nonprofit corporation, 23

O

Obama, Barack, 40
Ocrant article, 47
Office of Thrift Supervision, 31
Off the shelf automated systems, 171
Orwell, George, 97
Ottoman Empire, 19

P

Parmalat, 77
Patriot Act of 2001, 197
Pension holidays, 82
Perry, Michael, 33, 34, 113
Polakoff, Scott M., 32, 43
Ponzi, Charles, 17
Ponzi schemes, 17–26, 74
    Baptist Foundation of Arizona, 22, 23
    "bling bling" lifestyle, 24
    deception, 21
    foil for thievery, 22
    gambling on property prices, 24
    game-changing idea, 19
    greed, 25
    Ivar Kreuger of Kalmar, 18
    knee jerk cynicism, 25
    Kreuger inspired financial statements, 21
    Kreuger paper, 20
    liberty bonds, 20
    Match King, 19
    monopoly rights, 21
    nonprofit corporation, 23
    pyramid scheme, 21
    real estate, 23
    risk appetite, 17
    Roaring Twenties, 19
    robber barons, 20
    self-deception, 25

    speculative investments, 23
    stock market crash of 1929, 21
    suicide, 21
    sunbelt states, 23
    unreason, 25
    virtual potash, 17
Porter, Eleanor H., 12, 17
Posey, Bill, 47
Power, Michael, 90
Programming
    languages, names of, 156
    without programmers, 164
Proxy votes, 86
Pyramid scheme; *See* Ponzi scheme

Q

Quiet killers, 124
Quirks (spreadsheets), 156

R

RAC; *See* Risk assessment and control
RedEnvelope, 160
Regulators, 29–51
    conspiratorial regulator, 31–49
        adequately capitalized institutions, 40
        Alt-A loan market, 33
        apathetic regulator, 43–49
        Big Wing Theory, 38
        black eye, 49
        bogus appraisals, 35
        CAMELS composite rating, 39
        capitalism, 40
        capital ratio, 39
        civil servant "no speak," 45
        eye-opening stats, 34
        Federal Home Loan Bank System, 34
        held for sale portfolio, 38
        hot money, 38
        IndyMac, 33
        leadership *mettle*, 37
        monitoring of institutions, 32
        no doc procedure, 36
        Ocrant article, 47
        Office of Thrift Supervision, 31
        panic, 37

regulator capture, 39
run on the bank, 41
savings and loan crisis (1980s), 31
stunning indifference, 43
thrift industry, regulation of, 31
toxic assets, 38
traditional banking model, 31
underwriting practices, 36
zombie regulator, 42
ethology, 29
liar loans, 50
unreason abounds in places where it must not, 29–30
actions inducing risk, 30
bird experiment, 30
cardboard dummy, 29
environmental cue, 29
predator, 29
Request for Proposal (RFP), 188
Retirement savings, 74
Risk assessment and control (RAC), 56, 112
Risk Information Requirements Table (RIRT), 186
RiskMetrics™, 216
Risk System Requirements Documentation (RSRD), 186
Risk systems vendor; *See* Failure, understanding the nature of
Roaring Twenties, 19
Robber barons, 20, 73

S

San Diego County Employees Retirement Association (SDCERA), 61
Sarbanes-Oxley Act, 95, 108, 161, 197
Schumer, Charles, 41
Securities & Exchange Commission (SEC), 43, 44, 75, 85
Self-deceit, 98
Self-regulation, exclusive reliance on, 82
Shareholder activism, 86
Sin eater, 111
Skilling, Jeffrey, 98
Smith, Adam, 13, 15
Social loafing, 184

Specific risk, 58
Split-strike conversion strategy, 10
Spreadsheet City, 155–176
C++ language, 155
critical shut down, 155
day of critical shutdown, 155
fable, 156
how to bring spreadsheet risk under control, 162–163
incentives to learn, 163
inventory, 162
lack of guidelines, 163
operational spreadsheets, 162
how to minimize risk through formal testing, 169–171
application development, 169
execution testing, 169
spreadsheet declared error-free, 169
team inspection, 171
litmus test for spreadsheet risks, 162
off the shelf automated systems, 171
potential of compilable spreadsheets, 166–167
decompiler, 166
languages, 166
programming skills, 166
principles of spreadsheet engineering, 164–166
programming without discipline, 165
programming without programmers, 164
rigor, 166
spreadsheet engineering, 165
programming languages, names of, 156
questions, 171–172
quirks and bugs, 156
rules for superior spreadsheet design, 167–169
assumptions, 169
design, 167
format, 169
information isolation, 167
named cells and ranges, 168
reading, 169
spreadsheet hell, 157–158
caution, 158

end-user computing, 157
    models, 158
spreadsheet risk management, 170
understanding the nature of
        spreadsheet error, 163–164
    audit, 163
    impact, 164
    qualitative errors, 164
    quantitative errors, 164
    taxonomies, 164
why spreadsheet failure costs big
        time, 158–162
    CEO responsibility, 160
    incentives to learn, 163
    packages, 161
    regulators, 161
    rekeying of information, 159
    software engineered products,
        159
Standard & Poor's 100 price index, 10
Stickiness, 207
Stock market crash of 1929, 21
Strategic security director, 123–137
    American Dream, 125
    availability cascade, 125
    behavioral bias, 125
    biases, 123
    catastrophic events, 124
    chief risk officers, value added by,
        131–133
        corporate risk culture, 131
        effective brake, 131
        responsibility, 132
        vision, 132
    CRO mandate, 133–134
    disproportionate reporting, 124
    environmental hazard, 123
    events capturing attention, 124
    great work (*General and Industrial
        Management*), 126–127
        hazard risks, 127
        Internet security, 127
        strategic security director, 126
        traditional executive structure,
            127
    homicide, 124
    interbank funds, market for, 125
    quiet killers, 124
    rags to riches articles, 125

rise of Fayol's "strategic security
        director," 127–129
    value-added key points, 133
    Warren Buffet principle of risk
        management, 129–131
        advertisement, 129
        financial calamity, 129
        government assistance, 130
        interview, 129
        organizational fix, 131
        scenario, 130
        self-motivated managers, 130
Subprime mortgage, 15, 118
Sunbelt states, 23
Sweetheart deal, 113
Systematic risk, 59

T

Taurus, 180
Tinbergen, Niko, 29, 49
Torus, 128
Toxic assets, 38
Trade Practices Act, 79
TransAlta Corporation, 160
Truth, unpalatable, 53–70
    Capital Asset Pricing Model, 58,
        67–68
    emperor of risk, 57–58
        Amaranth Advisors, 57
        lawsuit, 58
        multistrategy hedge fund, 57
    frequently recounted tale, 53
    key points, 64–65
    perception and reality about risk
        management, 62–65
        Group of Thirty, 62
        trade-off, 63
        value-added risk management,
            63
    plan, 53
    professional discipline, 53
    questions, 65–66
    redundancy of financial risk
        management, 58–59
        Capital Asset Pricing Model, 58
        economic slowdown, 59
        empirical studies, 59
        rocket scientists, 59

specific or unsystematic risk, 58
systematic or market risk, 59
repo trader, 54
risk manager as "quivering dastard,"
    59–62
  bankruptcy, 59
  Chartered Financial Analyst
    designation, 60
  company-specific risks, 60
  leveraged positions, 62
  SDCERA, 60, 61
  unsystematic risk, 60
trading floor, 65
vulgar, but common, perception of
    risk management, 54–57
  allegory, 56
  despised risk manager, 54
  Great Recession, 55
  greed, 54
  largest financial implosion since
    Great Depression, 55
  moneymakers, 56
  satirical story, 56
  theory of mental illness, 55
  Wall Street Crash of 1929, 55
Tyco, 73, 85, 90

U

UBS, 74
Undertaking in Collective Investments
    and Transferable Securities
    (UCITS III), 207–208
Unpalatable truth about risk
    management; *See* Truth,
    unpalatable
Unreason, greed and, 3–27
  die-back fungus, 5
  Dutch golden age, end of, 7
  Dutch psyche, 6
  Flynn Effect, unreliability of, 7–12
    average intelligence of the world,
      8
    bull-market rush, 10
    gullibility, 8
    Intelligence Quotient, 8
    Madoff's victims, 11
    Ponzi scheme, 9

signature crook, 9
split-strike conversion strategy, 10
Glad Game, unintended consequence
    of, 12–17
  BASEL II guidelines, 15
  contempt of risk, 15
  delusions, 14
  flawed self-assessment, 15
  flippers, 16
  incentivized speculation, 16
  irony, 17
  Johari window, 14
  joke among mortgage brokers, 16
  liar's loan, 16
  mortgage backed securities, 15
  natural disasters, 14
  no money down mortgages, 16
  Pollyanna, 12, 16
  real estate "rags to riches" articles,
    16
  "regular Joe" investor, 16
  subprime mortgage borrowers, 15
  tax code, 16
  unreason, 16
  Yorkshire phrase, 16
panic, 7
Ponzi schemes, 17–26
  Baptist Foundation of Arizona, 22,
    23
  "bling bling" lifestyle, 24
  deception, 21
  foil for thievery, 22
  gambling on property prices, 24
  game-changing idea, 19
  greed, 25
  Ivar Kreuger of Kalmar, 18
  knee jerk cynicism, 25
  Kreuger inspired financial
    statements, 21
  Kreuger paper, 20
  liberty bonds, 20
  Match King, 19
  monopoly rights, 21
  nonprofit corporation, 23
  pyramid scheme, 21
  real estate, 23
  risk appetite, 17
  Roaring Twenties, 19
  robber barons, 20

self-deception, 25
speculative investments, 23
stock market crash of 1929, 21
suicide, 21
sunbelt states, 23
unreason, 25
virtual potash, 17
Rembrandt tulips, 4
Unsystematic risk, 58, 60

V

Value at Risk (VaR), 63, 201–217
explanation, 202–203
confidence level, 203
holding period, 203
market risk, 202
portfolio value, 203
profits, 202
joyous exclamation of Simons,
204–205
reputation of value at risk, 205
skepticism, 204
stampede toward VaR, 205
unreason, 204
permanent loss of capital, 201
questions, 213
tipping point, 205–208
Basle Accord, 207
derivatives trading desks, 206
expansion of VaR, 207
global dominance, 205
Law of the Few, 205
purveyors of doubt, 206
risk management process, 208
stickiness, 207
ways to measure risk, 202
what the rocket scientists may not
tell you, but you need to know,
208–212
accuracy of VaR, 210–212
bangs and flashes, 211
battleground, 211
bell-shaped curve, 209–210
corporate-wide VaR estimate, 211

exact imprecision, 210–212
fat-tail problem, 210
jargon, 208
latest risk management technique,
208
law of frequency of error, 210
perceived complexity of VaR, 209
underlying instruments, 209
Vendor risk information systems,
178–179
Vendor selection; *See* Failure,
understanding the nature of
Voluntary codes of conduct, 83

W

Wall Street Crash of 1929, 3, 55
Warren Buffet principle of risk
management, 129–131
advertisement, 129
financial calamity, 129
government assistance, 130
interview, 129
organizational fix, 131
scenario, 130
self-motivated managers, 130
Wheeler, McArthur, 14
Wilson, Kumanan, 124
Winkler, Charles H., 58
Wolf pack capitalism, 86–88
boardroom capitulation, 88
Connecticut Retirement Plans and
Trust Funds, 86
Euler's identity, 86
financial transparency, 86
hedge funds, 87
proxy votes, 86
shareholder activism, 86
Woochoong, Kim, 83
WorldCom, 73, 77, 85, 93

Z

Zeitgeist, 180
Zombie regulator, 42

For Product Safety Concerns and Information please contact our EU
representative GPSR@taylorandfrancis.com
Taylor & Francis Verlag GmbH, Kaufingerstraße 24, 80331 München, Germany

www.ingramcontent.com/pod-product-compliance
Ingram Content Group UK Ltd.
Pitfield, Milton Keynes, MK11 3LW, UK
UKHW021614240425
457818UK00018B/561

* 9 7 8 0 3 6 7 3 8 1 3 1 8 *